AF807131

Birkhäuser

Classic Texts in the Sciences

Series Editors

Jürgen Jost, Max Planck Institute for Mathematics in the Sciences, Leipzig, Germany

Armin Stock, Center for the History of Psychology, University of Würzburg, Würzburg, Germany

Classic Texts in the Sciences offers essential readings for anyone interested in the origin and roots of our present-day culture. Considering the fact that the sciences have significantly shaped our contemporary world view, this series not only provides the original texts but also extensive historical as well as scientific commentary, linking the classical texts to current developments. Classic Texts in the Sciences presents classic texts and their authors not only for specialists but for anyone interested in the background and the various facets of our civilization.

Jürgen Jost

Bernhard Riemann — On the Hypotheses Which Lie at the Bases of Geometry

Second Edition

Birkhäuser

Jürgen Jost
Max Planck Institute for Mathematics
in the Sciences
Leipzig, Germany

ISSN 2365-9963 ISSN 2365-9971 (electronic)
Classic Texts in the Sciences
ISBN 978-3-031-91488-1 ISBN 978-3-031-91489-8 (eBook)
https://doi.org/10.1007/978-3-031-91489-8

Mathematics Subject Classification: 53-XX, 01A35-60, 53-03

1st edition: © Springer International Publishing Switzerland 2016
2nd edition: © The Editor(s) (if applicable) and The Author(s), under exclusive license to Springer Nature Switzerland AG 2025

This work is subject to copyright. All rights are solely and exclusively licensed by the Publisher, whether the whole or part of the material is concerned, specifically the rights of translation, reprinting, reuse of illustrations, recitation, broadcasting, reproduction on microfilms or in any other physical way, and transmission or information storage and retrieval, electronic adaptation, computer software, or by similar or dissimilar methodology now known or hereafter developed.
The use of general descriptive names, registered names, trademarks, service marks, etc. in this publication does not imply, even in the absence of a specific statement, that such names are exempt from the relevant protective laws and regulations and therefore free for general use.
The publisher, the authors and the editors are safe to assume that the advice and information in this book are believed to be true and accurate at the date of publication. Neither the publisher nor the authors or the editors give a warranty, expressed or implied, with respect to the material contained herein or for any errors or omissions that may have been made. The publisher remains neutral with regard to jurisdictional claims in published maps and institutional affiliations.

This book is published under the imprint Birkhäuser, www.birkhauser-science.com by the registered company Springer Nature Switzerland AG
The registered company address is: Gewerbestrasse 11, 6330 Cham, Switzerland

If disposing of this product, please recycle the paper.

Preface

This could be the plot of a novel: The main character is a shy, sickly, young mathematician, living in poor conditions at a German university in the middle of the 19th century. He does not succeed in establishing a closer contact with the greatest mathematical luminary of his time. He is working toward his habilitation degree (a prerequisite for becoming a candidate for a professorship at a university in Germany). Part of the process is a colloquium. For such a habilitation colloquium, the candidate must propose three topics from which the faculty can choose. The first two topics are derived from the significant technical contributions that he has already made. He finds it difficult to decide what he should choose for his third topic, partly because he believes that, as usual, the faculty will select the first topic on the list anyway. As the third one, he offers a rather vague natural philosophical theme. To his surprise and consternation, the faculty chooses that one. Instead of now familiarizing himself with the state of art of the discipline and, in particular the really significant prior discovery that shook the entire field, he immerses himself in the work of a rather obscure philosopher. But his lecture then penetrates as deeply as never before into a field that had occupied and challenged the greatest thinkers of mankind since classical antiquity, and it even hints at the greatest discovery of the physics of the following century. Even the contribution of the superstar of German science, who had independently approached the same subject from a different point of view, faded into insignificance in comparison with the depth of insight of our young mathematician. Other famous scientists entered the stage with grotesque errors of judgment on the topic and content of the habilitation lecture after it had been published by a friend after the untimely death of our hero. Subsequent generations of mathematicians worked out the ideas outlined in the brief lecture and confirmed their full validity and soundness and extraordinary range and potential.

However, this is not a novel, because something similar did actually occur. We hope and trust that readers forgive the author certain exaggerations, and, of course, in the following pages, everything will be represented correctly. The young mathematician was Bernhard Riemann, and the lecture was entitled "Ueber die Hypothesen, welche der

Geometrie zu Grunde liegen" (On the hypotheses which lie at the bases of geometry). The mathematical genius was Carl Friedrich Gauss, the scientific superstar Hermann von Helmholtz, the fundamental prior mathematical discovery non-Euclidean geometry, the philosopher the now forgotten Johann Friedrich Herbart, the discovery in physics the theory of general relativity of Albert Einstein. The people who came up with completely wrong judgements included the psychologist Wilhelm Wundt and the philosopher Bertrand Russell. The friend who took care of the posthumous publication was the mathematician Richard Dedekind. The generations of subsequent mathematicians for whose research Riemann's ideas were a major inspiration include the author of these lines.

Typically, scientists read a scientific text from the point of view of the current state of science, interpret it in terms of subsequent developments, and seek at best unexplored potential for current scientific problems. Historians, however, want to determine the position of a text within the discourse of its time, reconstruct its origin and analyze the history of its reception. Although in the current debate on the role of the humanities, the importance of the historical sciences for understanding the present is emphasized, mathematicians are interested in the timeless content and not in the historical contingencies of scientific texts. Scientific projects that proved futile are either of no interest to the scientist or constitute annoying obstacles on a path that could have been straighter. For historians, in contrast, they can provide important insights into the history of ideas and the dynamics of discourses. For the scientist, texts whose effect has faded are without interest. For the historian this loss of interest is part of the reception history.

This edition of Riemann's "Ueber die Hypothesen, welche der Geometrie zugrunde liegen" tries to accept these challenges. As the editor is a professional scientist, the history is also sometimes read backwards. In particular, for this edition no thorough philological studies have been carried out. But he has also acquired some experience and expertise as a historian of science and tries to develop a historical perspective. He realizes, but perhaps does not always manage to escape the danger of assigning to terms and concepts their contemporary meaning and ignoring their historical context.

Although there will also be a formal mathematical chapter, I shall mainly attempt to explain the basic concepts and basic ideas in words, even if this will inevitably lead to some loss of precision.

While the first English version already contained a more detailed mathematical treatment than the original German one, for the second German version I have further expanded that mathematical treatment, and this then in turn has been used for the current second English version. Like the first edition, it provides the mathematical background and puts Riemann's reasoning into the more general and systematic perspective achieved by his followers on the basis of his seminal ideas. Readers who are not so much interested in mathematical details may skip that part, since I have also explained Riemann's reasoning verbally as an alternative to mathematical deductions. The description of contemporary research has also been expanded, in particular with a section on metric geometry and applications in machine learning. In fact, the current development of metric geometry is to a large extent based on extensions of Riemann's original concept of curvature, and that also is useful for interpreting discrete metric data.

I should repeat my gratitude to Erhard Scholz, Rüdiger Thiele and Klaus Volkert, for their very useful comments, corrections, suggestions and references on the first German version. Of course, the responsibility for any shortcomings rests with me alone. I am also grateful to the late Helmholtz expert Jochen Brüning for his insightful comments. Later, several discussions with Vincenzo De Risi helped me to understand the historical development of the concept of space and its philosophical implications more deeply. And I thank Domenico Giulini for referring me to an article of the theoretical physicist Markus Fierz, in which Newton's space concept and its historical roots are analyzed, as well as other helpful comments.

I thank Ingo Brüggemann, the librarian of the Max Planck Institute for Mathematics in the Sciences, and his staff for their valuable and efficient assistance in the acquisition of literature.

I owe my greatest debt of gratitude to my friend, the late Olaf Breidbach, for his initiative in establishing the series in which the original German edition of Riemann's text could appear, as well as for the many discussions we shared over several decades on a wide range of scientific topics. Together, we founded this series of Classic Texts in the Sciences and its German-language counterpart with great enthusiasm. I am pleased to say that it is now well established within the community, with many excellent contributions, several of which are also relevant to the present work. I hope that it will continue to preserve his spirit as one of the great historians of science of our time.

Leipzig, Germany — Jürgen Jost
December 2024

Competing Interests The author declares the following as potential competing interests: I am also an editor of the series in which this text will appear.

Contents

Introduction

1

A mathematical lecture without formulas, a geometric treatise without pictures or illustrations, a manuscript of only 16 pages that just came into being by chance, but a text that has shaped mathematics like few others works, which were all significantly longer, considerably more detailed, and much more carefully worked out. In this regard, we might mention the *Methodus inveniendi* by Leonhard Euler,[1] which founded the calculus of variations; Carl Friedrich Gauss' *Disquisitiones arithmeticae* that established mathematics as an independent discipline;[2] Georg Cantor's set theory, which introduced the modern conception of the infinite in mathematics; the theory of transformation groups of Sophus Lie, that is, the systematic study of symmetries that forms the mathematical basis for quantum mechanics; the programmatic writings of David Hilbert on the axiomatic foundation of various mathematical disciplines; or more recently the work of Alexander Grothendieck on the systematic unification of algebraic geometry and arithmetic. We are talking here of Bernhard Riemann's *Ueber die Hypothesen, welche der Geometrie zugrunde liegen* (*On the hypotheses which lie at the bases of geometry*), and this short script, written in 1854, but only published in 1868 after Riemann's death, whose wide ranging effects even take it beyond these works. This is because its position is at the intersection of mathematics, physics and philosophy, and it not only founds and establishes a central mathematical discipline, but also paves the way for the

[1] I am also preparing an edition of this text for the current series.

[2] In the sense that it develops its problems autonomously and intrinsically, instead of obtaining them from physics or other sciences.

© The Author(s), under exclusive license to Springer Nature Switzerland AG 2025

J. Jost, *Bernhard Riemann — On the Hypotheses Which Lie at the Bases of Geometry*,
Classic Texts in the Sciences, https://doi.org/10.1007/978-3-031-91489-8_1

physics of the twentieth century and at the same time represents a timeless refutation of certain philosophical conceptions of space. In the present volume, this key text of mathematics will be edited, positioned in the controversies of its time, and its effects on the development of mathematics will be analyzed and compared to those of its opponents. Riemann's *Ueber die Hypothesen, welche der Geometrie zugrunde liegen* has shaped and transformed mathematics in a manner very different from, say, Euclid's Elements, the writings of Leibniz and Newton on the creation of the infinitesimalsimal calculus or the above-mentioned works. It has, in a manner no less fundamental than those, influenced the development of mathematics as a science. Moreover, this text is essential for Einstein's theory of General Relativity. More recently, it also provided the mathematical structure underlying quantum field theory and its developments in theoretical Elementary Particle Physics (superstring theory, quantum gravity etc.).

However, the history of its influence and impact is not linear. Riemann's programmatic writing *Ueber die Hypothesen, welche der Geometrie zugrunde liegen* called to the scene the leading German physicist of his time, Hermann Helmholtz (later knighted, therefore von Helmholtz), the title of whose counter-essay *Über die Thatsachen, die der Geometrie zugrunde liegen* (*On the facts underlying geometry*) already pointed out a conflicting position and approach. (Though, in the text, the similarities with Riemann dominate[3] and its main thrust was not against Riemann, but against the Kantian concept of space.)[4] It would, however, be incorrect and misleading to view Helmholtz' work on the foundations of geometry simply as the by now obsolete and forgotten opposition of the established authority against the young genius, of the representative of a conservative scientific attitude against the protagonist of a novel scientific direction. Riemann's work was partly motivated by somewhat vague natural philosophical speculations—and in turn his text did have major implications for natural philosophy—, whereas the origin of Helmholtz' considerations laid

[3] Helmholtz says that he had developed the essential elements of his consideration before learning of Riemann's work (which had been published with a 14 year delay), but certainly later than Riemann, who had his lecture delivered and the script written in 1854.

[4] It might appear natural to contrast this work here with Riemann's text. After giving this option serious thought, however, in the end I refrained from it, because this work of Helmholtz did not achieve the same depth and elegance as Riemann's. Moreover, among the various writings of Helmholtz on epistemological issues, this particular text is not the best and the clearest, and so, by the choice of that particular work, the important physiologist and physicist Helmholtz would have appeared in a wrong light. So if we had wanted to represent Helmholtz' theory here by one of his writings, then we should have selected another of his writings, namely *Über den Ursprung und die Bedeutung der geometrischen Axiome* (*On the origin and the importance of the geometrical axioms*) or his inaugural address as Rektor (president) of the University of Bonn *Die Tatsachen in der Wahrnehmung* (*The facts in perception*), but then we would have lost the close relationship with Riemann's habilitation address.

in sensory physiology—and his ideas remain highly relevant here. Moreover, Helmholtz influenced another fundamental mathematical theory, the theory of symmetry groups of Sophus Lie. Although Lie sharply criticized the mathematical aspects of Helmholtz' work, he nevertheless adopted the latter's conceptual approach. Lie's theory of symmetry groups has become one of the cornerstones of quantum mechanics. The concepts of symmetry and invariance link the intuition of modern physics with the mathematical frame of geometry in the sense of Riemann and Einstein. In this sense, Helmholtz' text contained an aspect that was visionary for modern physics, even though this became clear only through the work of Lie and probably was quite different from what Helmholtz himself may have had in mind.

Riemann was probably motivated by somewhat vague nature-philosophical speculations—and his work then conversely had significant implications for the philosophy of nature. In contrast, Helmholtz's considerations firstly had their starting point in sensory physiology—and his ideas remain relevant here—and secondly, they also influenced a very important mathematical direction, namely the theory of symmetry groups of Sophus Lie. Even though Lie sharply criticized the mathematical details of Helmholtz' treatises, he took over the latter's conceptual approach. Lie's theory of symmetry groups became one of the essential foundations of quantum mechanics, and the concepts of symmetry and invariance connect the physical intuition of modern physics with the mathematical framework of geometry in the sense of Riemann and Einstein. In this sense, also Helmholtz' text played a pioneering role for modern physics. In contrast to Riemann's text, that text exerted its influence not directly, but only through the work of Lie, and probably this influence was rather different from what Helmholtz himself had imagined.

It is noteworthy that Riemann's *Hypotheses* as one of the key texts in mathematics proceeds without mathematical formulas (in the whole text, there is only a single formula which is of only marginal importance). This sets Riemann's text apart from other foundational mathematical works, like the sophisticated and deeply thought out symbolism of Leibniz or the formalization of the infinite of Cantor. Even his most important precursor, Gauss' *Disquisitiones generales circa superficies curvas*, which founded modern differential geometry, the starting point of Riemannian Geometry, is different in this respect. At least in this case, the history of mathematics is not simply a progressive formalization, but it turns out that mathematical abstraction can in principle rise well above formulas .[5]

[5] Of course, the occasion of Riemann's text should be taken into account. It was a colloquium for the Faculty of Philosophy, and Riemann certainly wanted to respect the lack of mathematical knowledge of most of the people in his audience. Among these, besides Gauss, who was not a professor of mathematics, but a professor of astronomy and director of the observatory, mathematics was represented only by the two Professors Ulrich (1798–1879) and Stern (1807–1894). However, other such lectures or texts, like Klein's Erlangen program with which he introduced himself to the faculty in Erlangen, certainly could be much more formalized, and if the faculty had chosen one of the other topics suggested by Riemann, the presentation would probably have been developed in mathematical formulas.

Bernhard Riemann has decisively shaped modern mathematics to such a degree that only the influence of Carl Friedrich Gauss is comparable with his. Not only did he found modern geometry with his habilitation lecture published here—and the most important part of modern geometry is therefore called Riemannian geometry—, but he has created a number of basic theories and introduced many fundamental concepts that guided and influenced many other areas of mathematics. His concept of a Riemann surface combined in an ingenious manner complex analysis and the theory of elliptic integrals. This work was at the same time the starting point for the development of topology, i.e., the investigation of forms and shapes independent of metric properties, in contrast to Riemannian geometry. It also had a decisive impact on modern algebraic geometry. On top of that, it introduced completely novel analytical tools in the theory of functions of a complex variable. The latter, even if initially Weierstrass detected and pointed out essential analytical gaps that could be closed only later by Hilbert, paved the way for the modern calculus of variations and the existence theory for solutions of partial differential equations. Those in turn, as implemented and controlled with the methods of numerical analysis, constitute a fundamental tool of modern engineering. A novel and pathbreaking idea was that Riemann no longer tried to approach analytic functions in the complex plane through an analytical expression, but rather considered them as determined by their singularities (poles, i.e., points where they become infinite, or branch points). In this way he could assign to such a function a so-called Riemann surface and then determine the qualitative properties of the function in terms of the topology of that Riemann surface. This radiated in almost all areas of modern mathematics, and for example, even revolutionized number theory, the analytical expressions of which could then also be interpreted and treated by geometric methods. It was likewise a pathbreaking aspect of the theory of Riemann surfaces that Riemann not only looked at a single mathematical object, but conceptualized a class of objects through the variability of parameters. This led to the theory of moduli spaces which is basic for algebraic geometry. For this reason, Riemann surfaces also constitute the basic objects of the currently perhaps most promising theory, string theory, for the unification of the known physical forces. The so-called Riemann-Roch theorem [Gustav Roch (1839–1866) was a student of Riemann, who completed Riemann's work on these issues] was one of the guiding principles of mathematics in the second half of the twentieth century and resulted in the works of Hirzebruch, Atiyah-Singer and Grothendieck that produced key results of modern mathematics. The Riemann hypothesis, more than 150 years after its formulation, is still considered as the hardest and deepest open problem of all of mathematics.

Print by August Weger, after a photograph of Riemann, probably taken in 1863

On Riemann's biography: Bernhard Riemann, the son of a Lower Saxon protestant minister, lived from 1826 to 1866. He remained very attached to his family. Several early deaths put the family into a difficult financial position that led to unsecured financial circumstances, and Riemann had to struggle hard to support his three surviving sisters. Like most of the great of the history of mathematics, he showed an extraordinary mathematical talent already as a schoolboy. After some hesitation, he followed this talent and studied mathematics instead of theology as wished by his father, in the scientific centers Göttingen and Berlin. His main academic teachers and role models were Carl Friedrich Gauss (1777–1855),[6] with whom he received his doctorate in 1851, and Peter Gustav Lejeune Dirichlet

[6] Gauss was born in Brunswick in modest circumstances. Since his outstanding mathematical talent was recognized early on, he was, however, generously supported by the Brunswick Duke. Already at a young age he made significant mathematical discoveries, such as on the question of the constructability of regular polygons. His *Disquisitiones Arithmeticae*, published in 1801, but written already some years earlier, are considered as the work that founded modern mathematics as an autonomous science. A spectacular success of his mathematical methods of error estimates was the rediscovery of the minor planet (asteroid) Ceres in the same year. This minor planet had been discovered by astronomers, but then again lost sight of until the Gaussian methods of path calculation would permit prediction of its position with high enough precision so that the astronomers knew to which position in the sky they had to turn their telescopes to find it. Since 1807, Gauss was a professor in Göttingen and the director of the observatory. Gauss is considered the greatest mathematician of all time, and he has

(1805–1859),[7] of whom he attended many lectures in Berlin. Dirichlet in 1855 became the successor of Gauss in Göttingen, and in 1859, Riemann in turn became his successor as a full professor in Göttingen after he had been appointed in 1857 as an associate professor (a so-called Extraordinarius that did not receive a regular salary but could only collect fees from the students attending his lectures). He was shy and sickly, but impressed the scientific world by the richness of his mathematical insight and the boldness and originality of his mathematical theories.

When studying in Berlin, he had discussions with the somewhat older, but short-lived mathematician Gotthold Eisenstein (1823–1852), but both were too shy to enter into a closer personal relationship. When studying in Göttingen, he was an active participant in the seminar of the physicist Wilhelm Weber (1804–1891), which inspired his search for a unified physical field theory.[8] He developed closer personal contacts outside of his family only with

---

influenced almost all areas of modern mathematics and even founded many of them. Together with the physicist Wilhelm Weber (1804–1891) he constructed the first telegraph. The mathematical methods developed by him are fundamental for astronomy and geodesy. Especially in old age, Gauss was difficult to approach, undoubtedly also due to a not very happy family life, and the shy Riemann could not establish direct personal contact with him. Riemann therefore acquired the mathematical theories and discoveries of Gauss by self-study.—A recent biography of Gauss is Kaufmann-Bühler (2012).

[7] Dirichlet was born in Düren in the Rhineland as a son of the local postmaster, whose father had immigrated from the Walloon region in present-day Belgium, hence the Romanesque name. During a stay in Paris from 1822–1827, as a foreigner he was not allowed to attend the courses of the then leading French mathematician Augustin Louis Cauchy (1789–1857) at the Ecole Polytechnique. Fortunately, he succeeded in gaining access to the circles of Jean-Baptiste Louis Fourier (1768–1830), who, starting from physical problems of thermodynamics, had introduced the famous series representations for periodic functions. Dirichlet proves a basic result about these series expansions. Alexander von Humboldt (1769–1859), who after his famous expeditions initially stayed in Paris and then held influential positions in Berlin, is impressed by him and supports and encourages him and brings him as a professor to Prussia, first to Breslau and then in 1829 to Berlin. Dirichlet and his friend and colleague Carl Gustav Jacob Jacobi (1804–1851) turn the University of Berlin, which had been founded in 1810 by Wilhelm von Humboldt (1767–1835) in the course of the reforms motivated and necessitated by the Napoleonic aggression against Prussia, into a center of mathematical research. Dirichlet's wife Rebecca was highly educated; she came from one of the leading Jewish families in Germany that also played a key role for the emancipation of the European Jews. She was a granddaughter of the philosopher Moses Mendelssohn (1729–1786), a niece of the author Dorothea (von) Schlegel (1764–1839), who in turn was the wife of the writer and theorist of Romanticism Friedrich (von) Schlegel (1772–1829), and a sister of the composer Felix Mendelssohn Bartholdy (1809–1847), who as head of the Leipzig Gewandhaus Orchestra, initiated the rediscovery and renaissance of the baroque music of Bach and Händel. In this way, Dirichlet's life was intertwined with those of many other prominent personalities. Dirichlet was friendly and open towards Riemann, and Riemann could learn a lot from him. Dirichlet made in particular important contributions in number theory, and he founded the analytic direction of number theory. A historically oriented introduction can be found in Scharlau and Opolka (2010). The principles applied by Dirichlet in the calculation of variations later played a central role in Riemann's studies on function theory and Riemann surfaces.

[8] S.z.B. Jungnickel and McCormmach (2017).

the younger mathematician Richard Dedekind (1831–1916).[9] He went through the steps of a standard academic career from the lectureship to a professorship in Göttingen. The salary that came with this professorship eased his financial situation considerably, in particular because after the death of his parents and his brother he also took over the responsibility for three unmarried sisters. Health problems necessitated interruptions of this position by extended sojourns in Italy whose climate was more suitable for him, but where he succumbed to lung disease before reaching the age of 40, leaving behind his wife and a young daughter.

Riemann died neither as young as Niels Hendrik Abel (1802–1829) nor Evariste Galois (1811–1832), who in their short lives could create only one important mathematical theory (that of the Abelian integrals and group theory), nor did he reach the old age of the often grumpy and withdrawn Gauss. He had neither the almost inexhaustible vitality of Leonhard Euler (1707–1783) nor the active energy of Carl Gustav Jacob Jacobi (1804–1851) and Felix Klein (1849–1925). He could not rely on a group of talented young students and collaborators as could David Hilbert (1862–1943), for the requisite institutional conditions were established in Germany only later by Felix Klein and others (and then destroyed again by the Nazis through expulsion and murder of Jewish mathematicians and the exile of those who were not Jewish but only dissenting). But Gauss and Riemann created the rise of mathematics in Germany and especially in Göttingen, that made such an institutionalization possible.

As far as the author knows, there is no detailed biography of Riemann written for general readers.[10] Otherwise, biographies of prominent mathematicians are not rare and in some

[9] On Dedekind see Winfried Scharlau (ed.), *Richard Dedekind. 1831|1981*, Braunschweig/Wiesbaden, Vieweg, 1981. The letters printed there also contain biographical material on Riemann, which can complement the picture in Dedekind's biography of Riemann in the latter's collected works.

[10] In Riemann's collected works edited by Heinrich Weber and Richard Dedekind, there is a a twenty page biography of Riemann, written by his friend and colleague Dedekind. Hans Freudenthal wrote a short biography for the Dictionary of Scientific Biography. In addition to other short biographical sketches, there are the scientific biographies of Michael Monastyrsky (1998) and Detlef Laugwitz (2008) in which the development and impact of Riemann's scientific work is placed in the context of the circumstances of his life. The scientific biography of Laugwitz has been of great help to me at various places, and it also contains a detailed description of Riemann's life that is accessible to a general readership. Various other such analyses can be found in the reissue of Riemann's collected works edited by Raghavan Narasimhan. A systematic research of the unpublished scientific notes and sketches and the available biographical material on Riemann has been started by Erwin Neuenschwander. For some results, see Neuenschwander (1996). Various mathematical historical studies discuss the development of geometry before, through and after Riemann, but usually not from a biographical perspective. Sources can be found in the bibliography at the end of this book.

countries constitute even an expression of national pride, like the biographies of the Norwegian mathematicians Niels Hendrik Abel and Sophus Lie (1842–1899) by Arild Stubhaug. In other cases, such as the biographies of David Hilbert and Richard Courant (1888–1972) by Constance Reid that are very popular among mathematicians, their lives under the often difficult and unfavorable circumstances of their times and the historical events affecting them also arouse interest. Since Riemann's life unfolded in a quiet time without dramatic personal or historical events, there is no material for a spectacular tale. Also, the cult of genius, for which Riemann should actually provide an excellent example, did not take him up as it had the younger mathematicians Abel and Galois or the artists Raphael, Mozart and Schiller who had a life span similar to that of Riemann.

The academic life of Bernhard Riemann appears unproblematic from today's perspective if we forget the precarious financial circumstances in which the so-called privatdozenten (private lecturers) had to live at that time. These privatdozenten were scientists qualified by a doctorate or a habilitation, but who did not have a regular professorship. The promotion from a privatdozent to a professor was probably already in his time considered as the usual academic career. It should be pointed out, however, that in many cases there were deviations from this path, in both the positive and the negative sense. In any case, the modern university system had been founded by Wilhelm von Humboldt only half a century before Riemann, and the relocation of scientific research from the scientific academies and learned societies of the 18th to the universities of the 19th century and the establishment of appropriate academic career structures had needed some time. In particular, in the initial phase, the university system could therefore typically not train its next generation through internal university career paths, but had to recruit university teachers often from outside, from the group of high-school professors or that of the scientific practitioners, who worked in astronomical observatories, botanical gardens, pharmacies or other institutions. Conversely, aspiring scientists could therefore not necessarily build a purely academic career, but often had to take long biographical detours. Thus, on one side, there were gifted and talented scientists who never in their lives got a university position. On the other side, there were those who succeeded in gaining entry into an academic career from an outsider position, or conversely, those who from an early age were generously supported by noble sponsors or governments. The latter category included Gauss, who was supported by his ruler, the Duke of Brunswick, or Dirichlet, who at the initiative of Alexander von Humboldt was first sponsored during his studies in Paris and then given a professorship in Berlin. A well-known example outside of mathematics is the chemist Justus (von) Liebig (1803–1873), for whose studies in Paris the Grand Duke of Hesse provided and who then, as early as 1824, was made a professor at Giessen. There Liebig established the institution of the chemical laboratory for teaching and research. Notable examples of those, who through tough and sustained efforts managed external access to and then achieved central positions in the German scientific system, are the mathematician Karl Weierstrass (1815–1897), who had to spend many years as a grammar school teacher in Deutsch-Krone in former West Prussia and in Braunsberg in former East Prussia before he could gain scientific recognition by his mathematical work on elliptic integrals, or Hermann (von) Helmholtz (1821–1894), who first had to work as a military surgeon before he was able to start his academic career. Wilhelm Killing (1847–1923) conducted his important studies on the foundations of geometry and the infinitesimal transformation groups (Lie algebras) in the little time left by a huge teaching schedule that comprised all sciences and on top of which he even had to serve for some periods as the principal at the Lyceum Hosianum in Braunsberg in East Prussia, where his teacher Weierstrass had worked before. In 1892, Killing became a professor in Münster. There, however, his teaching and administrative tasks up to the office of the Rektor (University President) and his caritative commitments that were rooted in his catholicism took so much of his time that he could barely continue his mathematical research. Others, such as the mathematician Hermann Grassmann (1809–1877), were denied scientific recognition throughout their lives. Grassmann was a high school teacher in Stettin, and he founded linear algebra, which is fundamental in today's mathematical university education and is taught from the very first semester. (Grassmann was also a major Sanskrit researcher and studied in particular the Rigveda. In contrast to his mathematical work, these studies obtained the recognition of the academic world.) A well-known example outside of mathematics is of course the Augustinian monk Gregor Mendel (1822–1884), whose discovery of the quantitative laws of inheritance, one of the deepest insights throughout the history of biology, failed to gain the notice of the professional biologists, until after the turn of the century they were rediscovered by several researchers, in a weaker form first, and then became the foundation of the modern gene concept.

Previous editions (for details see the bibliography at the end):

- Collected Works (various editions, the most recent by Narasimhan);
- Weyl, with a detailed mathematical commentary.

This plaque in the cemetery at Biganzolo on Lake Maggiore commemorates Riemann. (Photo by G. Rondena—personal photo, Public Domain, https://commons.wikimedia.org/w/index.php?curid=8374862)

References

Jungnickel, Christa and Russell McCormmach. 2017. *The Second Physicist. On the History of Theoretical Physics in Germany*. New York: Springer.

Kaufmann-Bühler, Walter. 1981. *Gauss: a biographical study*. New York: Springer.

Laugwitz, Detlef. 2008. *Bernhard Riemann 1826–1866. Turning points in the conception of mathematics*. Boston: Birkhauser.

Monastyrsky, Michael I. 1998. *Riemann, Topology, and Physics*. New York: Springer.

Neuenschwander, Erwin. 1996. Riemanns Einführung in die Funktionentheorie: eine quellenkritische Edition seiner Vorlesungen mit einer Bibliographie zur Wirkungsgeschichte der Riemannschen Funktionentheorie. *Abhandlungen der Akademie der Wissenschaften zu Göttingen, Math.-Phys. Klasse*, p. 44

Scharlau, Winfried and Hans Opolka. 2010. *From Fermat to Minkowski. Lectures on the Theory of Numbers and its Historical Development*. New York: Springer.

Historical Introduction 2

2.1 The Space Problem in Physics, from Plato and Aristotle to Newton and Leibniz

Riemann's text links in a novel manner different thematic threads that combine mathematics, physics and philosophy, and Helmholtz in addition brings sensory physiology into the discussion. Therefore, in order to historically situate Riemann's text, let us first sketch the history of the space problem as developed in those sciences. The starting point of geometrical research is Euclid (fl. 300 BC). As is well known, in his Elements, from a few definitions, postulates and axioms, he developed a planar and spatial geometry in a constructive manner. This then dominated the subsequent development of geometry so strongly that often and for a long time, it was considered as being without alternatives.

However, Euclidean geometry did not arise out of nothing, but in a rich philosophical and mathematical context, which was essentially shaped by the theories of Plato(428/7–348 BC) and Aristotle (384–322 BC). And we begin our presentation with Plato, whose essential writings have been handed down in their entirety, while the writings of his predecessors have only survived in fragments and are difficult to reconstruct.

Plato sets out his conception of space in the *Timaeus*, embedded in general considerations on the origin and structure of the cosmos. He says that being, space and becoming already existed before the cosmos, and it is disputed among Plato scholars to what extent this "before" is meant temporally or ontologically. The space ($\chi\acute{\omega}\rho\alpha$, chōra) is considered as amorphous, shapeless, and can therefore receive arbitrary objects. However, such arbitrary objects are composed of the elements fire, water, air and earth. And these in turn consist of combinations of geometric structures. The basic elements are two right-angled triangles. The first of these triangles is isosceles, and from four of these triangles a square can be formed.

© The Author(s), under exclusive license to Springer Nature Switzerland AG 2025

J. Jost, *Bernhard Riemann — On the Hypotheses Which Lie at the Bases of Geometry*, Classic Texts in the Sciences, https://doi.org/10.1007/978-3-031-91489-8_2

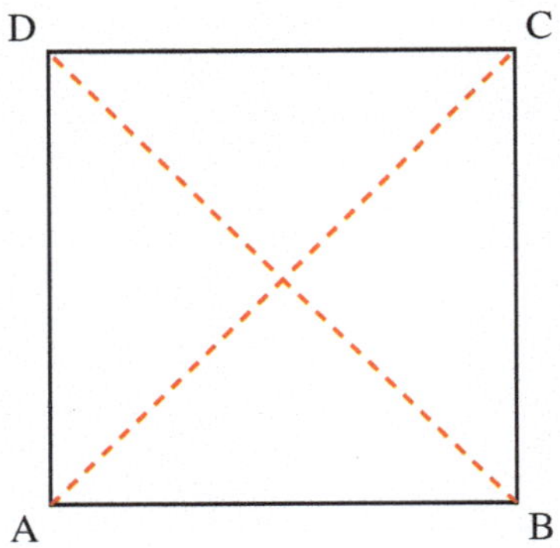

The other right-angled has sides of different lengths, and their ratios are such that six of them can be used to form an equilateral triangle.

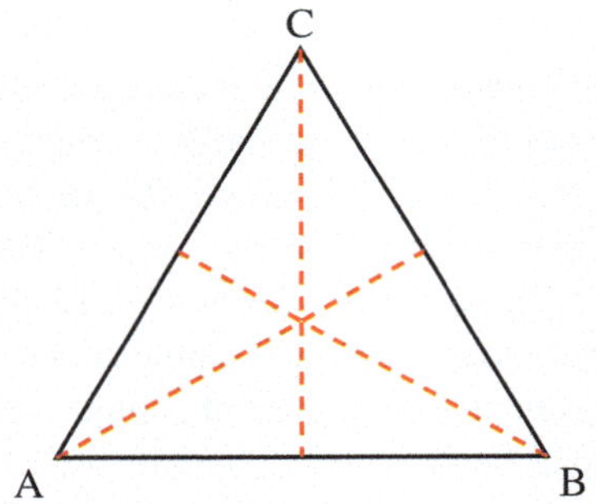

If we use the Pythagorean theorem already known to Plato and consider the side lengths of the equilateral triangle a, the cathets are $\frac{a}{2}$ and $\frac{a}{2\sqrt{3}}$, while the hypotenuse is $\frac{a}{\sqrt{3}}$. However, the unit of measurement a is undetermined and only the side ratios are are determined. Thus, one of the cathets is the half of the hypotenuse.

These surfaces, i.e. the square and the equilateral triangle, in turn form the elementary components from which four of the Platonic solids, cube, tetrahedron, octahedron and icosahedron, are composed. The cube then corresponds to earth, the tetrahedron to fire, the octahedron to air and the icosahedron to water. The fifth Platonic, the dodecahedron, body is assigned to the creator of the world, the demiurge, or the world itself as a whole. The cube, tetrahedron and dodecahedron were already already known to the Pythogorians. The octahedron and the icosahedron were discovered by the mathematician Theaitet, a friend of Plato. Theaitet probably also proved that these five were the only regular bodies. An echo of such constructions can still be found in the speculation of the early Kepler, who described the orbits of the six planets known at the time in the Copernican system, Mercury, Venus, Earth, Mars, Jupiter and Saturn, by means of such bodies, before he later, after years of laborious computations on the astronomical data obtained by Tycho Brahe, arrived at his fundamental planetary laws.

The four elements earth, fire. air and water had been introduced by Empedocles (ca. 500–430 BC) as basic elements and formed a fixed building block of Greek thought. The real bodies then consist of a nesting of such ideal bodies of different sizes. And movement arises from the fact that these change their sizes and because of their nesting then also their location, and the compositions thereby change their state. Although this probably originated from the ideas of the atomists around Leucippus (mid fifth century BC) and Democritus (ca.460–370 B.C.), in the end it is very different from these, according to which small particles that cannot be further decomposed, the atoms, move in an otherwise empty space. These atoms possess gravity and resistance and therefore can be distinguished by their shape, position and arrangement, and can hence combine with each other in many different ways.

Although these Platonic ideas are imaginative, they may seem strange to us. But they contain an idea that is still virulent today, namely that spatial matter is ultimately made up of basic building blocks that are no longer spatial, in this respect similar to the strings of modern elementary particle physics. Leibniz will later propose ultimate constituents of matter, the monads, that are no longer spatially conceived, but which unlike the Platonic triangles, are not homogeneous but individually unique.

Another aspect of the Platonic theory is that the basic building blocks, i.e. the right-angled triangles, do not have a fixed intrinsic size, but can be realized in different sizes. The quantitative dimensions of spatial objects therefore are only secondary properties.

Plato imagines the universe to be spherical, and the striving towards its center causes the gravity of the bodies. As a sphere, the universe is isotropic, i.e. all directions emanating from the center are equal, and all points on the periphery are geometrically equal.

For Plato, the physical and sensually perceptible world is therefore ultimately not material, but geometrically constituted.

The space concept from antiquity that became influential and fundamental, in fact dominant, was that of Euclid, or more precisely that extracted from Euclid's *Elements* (Heath 2000), even though never explicitely stated there. And in this sense, we shall speak of Euclidean space.

While the relationship of Euclidean geometry to the philosophy of Plato was unproblematic, it did not fit with Aristotelian physics. Euclidean space is homogeneous, that is, any point in it is like any other, and isotropic, that is, in all directions it looks the same. No point and no direction is in any way distinguished. Aristotle (384–322 BC) in contrast thought of the world as a collection of places.[1] According to him, the location of an object

[1] While originally the two expressions χώρα, chōra, and τόπος, tópos, did not have clearly differentiated meanings, chōra is usually understood as *space*, in the sense of something extended, and tópos is translated as place, location, position. An Aristotelian place is a topos. Among the atomists and other ancient thinkers also τὸ κενὸν, to kenon, the void, was used. For details cf. Algra (1994, p. 34ff).

is determined by its bounding surface. Every body has its natural place to which it tries to move. Thus, the world is heterogeneous. Because objects naturally fall downwards, the vertical direction is different from other directions. Thus, Aristotelian space is not isotropic. Contrasting Euclid and Aristotle in this manner leads us already to the fundamental question of the relationship between geometry and physics, or in a slightly different formulation, to the question about the relation between the geometric space and the objects filling it. From the point of view of physics, this also raises the question about the existence of the vacuum, the empty space devoid of any content that was required for the ancient atomic theory of Democritus (ca. 460–370 BC) and Leucippus (fifth century BC), but which was considered impossible by Parmenides and Aristotle. Euclidean space is infinite,[2] and the question of the finiteness or infinity of physical space was also controversial in antiquity, with Aristotle again standing on the opposing side. For him, infinity could just exist as potentiality in time, but not actually in space.

A new point was then introduced by the artists and art theorists of the Italian Renaissance. They wanted to represent objects no longer in their real, objective size or display persons with size corresponding to their significance, but show them as they presented themselves subjectively to the eye of the beholder. For this purpose, they had to utilize the objectively valid laws of geometrical optics, which in turn follow the rules of Euclidean geometry. In a certain sense, they replaced the physics of bodies by a physics of light rays which had to correspond to Euclidean geometry.[3] This may also have been facilitated or inspired by the needs of cartography required for the rise of maritime trade, which was also concerned with the adequate representation of spatial relations.[4] Anyway, linear perspective, which is said to have been discovered by the Florentine architect and artist Filippo Brunelleschi (1377–1466) and which found its first representation in the book *Della Pittura* (1435) by the writer and scholar Leon Battista Alberti (1404–1472), is the Euclidean construction of the projection from the three-dimensional space on a two-dimensional surface. This inspired then Kepler (1571–1630) and Desargues (1591–1661) to a new treatment of conic sections.[5] In the hands of mathematicians, this led (only) in the first half of the nineteenth century to the development of projective geometry,[6] which then in connection with the ideas developed by Riemann, Klein and other mathematicians of the second half of the nineteenth century became a part of algebraic geometry.

---

[2] The concept of infinity in antiquity, however, differed from the modern one which was shaped by Cantor's view of modern mathematics. The infinite was understood not as actually existing, but as a potentiality, in the constructive sense that, for example, a straight line can be extended forever, without reaching an end, but without having to assign to all points of this infinite line a prior existence. For a systematic analysis of the historical development of the concept of infinity s. Cohn (1896).

[3] An influential treatise is Panofsky (1991, 1927).

[4] See Edgerton (2009).

[5] See for instance Field (1997).

[6] See the detailed presentation by Andersen (2008).

The Italian natural philosophy of the sixteenth century also began to propose alternatives to the hitherto dominant Aristotelian-scholastic worldview.[7] Julius Caesar Scaliger (1484–1558) revives the doctrine of the empty space of the ancient atomism, the precondition for space to become the container of objects. In contrast to Aristotle, the location of an object is then no longer determined by its bounding surface, but becomes the three-dimensional geometric content enclosed by such a boundary. Space thus no longer surrounds objects, but objects fill space. Bernardino Telesio (1508/9-1588) developed a dynamic anti-Aristotelian natural philosophy. For him, the (empty) space is incorporeal and ineffective; it can merely receive objects.[8] For Francesco Patrizi (1529–1597) space is origin and source of quantity and it constitutes the basis of the world of things.[9] Because it shows no resistance, it is not physical, but at the same time distinguished from purely spiritual entities by the feature of extension. Thus, space is here, in contrast to Aristotle, not a reality inherent in the objects, but independent of them. The ideas just discussed remained essential and effective for the further development of the concept of space.

The physics of Galileo Galilei (1564–1642), which establishes quantitative mathematical laws in contrast to the qualitative, logical reasoning of Aristotle,[10] assumes Euclidean

---

[7] For a systematic presentation of the entire historical development, we refer to Cassirer (1957) and the volume (De Risi 2015) edited by V. De Risi. A concise introduction to the development of anti-Aristotelianism and mechanical philosophy until Newton can also be found in Garber (2006).

[8] In addition to Cassirer, Vol. 1, loc .cit., see also the article on Telesio in Eisler (1912, S. 741f.).

[9] see the profound analyis in Fierz (1954) and De Risi (2016).

[10] Galileo put the focus on the empirically measurable development of physical processes instead of their derivation from final principles. He believed that the world therefore cannot be readily deduced from revealed principles, but can only be painstakingly explored empirically and needs to be measured. With these views and the underlying atomistic conceptions, Galileo ultimately demolished the scholastic natural philosophy of the Middle Ages that had been formed through the reception of Aristotle. According to scholastic philosophy, the world was a structure with an order designed for and revealed to man. (See, e.g., the concise analysis of Burtt (1932).) This philosophy also provided a foundation for the Aristotelian distinction between form and substance which in turn was needed for the doctrine of the Eucharist. This was an important component of the worldview of the Catholic Church that had hardened in the Counter-Reformation. This was a basic reason for opposition of the Church against the Galilean conception of nature. This is in contrast to the beginning of the dissolution of the Aristotelian world view in the Italian natural philosophy of the sixteenth century, as outlined above, which had still met with papal benevolence. Instead of the Aristotelian substances, which could be given different forms, and which then in turn could be converted while preserving their form in the miracle of the transubstantiation, in the Galilean view, there were only formless atoms whose qualitative properties such as color were constituted only in the process of perception. Then such a miracle became impossible, or plausible, at best, as a crude sensory illusion. And of course, also the Copernican heliocentric system did not harmonize with a human-oriented plan of creation. These seem to be the deeper underlying reasons for the resistance that Galileo encountered from the leading intellectual representatives of the Catholic Church. This is different from popular expositions, where petty disputes about the literal interpretation of certain biblical passages, like that where Joshua allegedly made the sun stand still when taking Jericho, are presented as the reason for the persecution of Galileo by the church. Bible passages could well be interpreted allegorically also by the Catholic Church, if this was deemed useful or necessary for systematic reasons. Scriptures

geometry. He considered idealized situations, like those of a ball rolling on an infinitely extended inclined plane or the uniform acceleration-free movement in empty space, which can be described mathematically in exact terms and at the same time approximates the physical processes in the real world. The difference between the ideal and the real movement is due (as opposed to Aristotle) to conceptually secondary effects such as friction or drag. The uniformity of the idealized physical processes presupposes the uniformity of the space in which it takes place. In modern terminology, the invariances of physical movements can be reduced to transformations of space that do not change its geometry. This is the so-called concept of Galilean invariance, that the laws of physics are the same in all reference frames that move relatively to each other at a uniform speed, without acceleration. This remains valid in Einstein's special theory of relativity, in which, however, the Galilei transformations are replaced by the relativistic Lorentz transformations, in which not only the spatial positions, but also time is transformed linearly. Such Lorentz transformations therefore no longer take place in three-dimensional Euclidean Space, but in a space extended by the inclusion of time, the four-dimensional Minkowski space.

So Galileo replaced the Aristotelian conception of an ordered and structured cosmos by the uniform laws of a per se unstructured universe (already by Giordano Bruno (1548–1600) enthusiastically propagated as infinite).[11] This was not only the crucial breakthrough of modern physics, but it also established the systematic questions and problems of modern geometry, which then reached their culmination in the work of Riemann.

For the physics of Isaac Newton (1642–1727), Euclidean space was the invariant container in which the physical objects, typically idealized as point masses, moved under the influence of forces. This pioneering concept had, however, to battle against the Cartesian idea that the characteristic criterion of matter was their extension;[12] to René Descartes (1596–1650), the Newtonian concept of a mass point would have appeared completely meaningless. Johannes Kepler prepared and Newton developed the notion of bodies that are not characterized by their spatial extension,[13] but rather by their dynamic force effects, that is, by space

probably served rather as material for rhetorical tricks at a time when the methods advocated by Machiavelli could also be employed in intellectual discussions. Even with physical experiments it is often unclear whether they were actually carried out or their results were only claimed on the basis of intuitive plausibility as evidence for a systematic theory, see e.g. Koyré (1937). Redondi (1987, 1983) collected evidence that the church actually saw justification of the doctrine of transsubstantion, which was central for the Counter-Reformation, endangered by Galileo and reprimanded him for this.

[11] The classical treatment is Koyré (1957). Cassirer (1957) contains more material and in some regards penetrates more deeply into the matter.

[12] For a modern account of Cartesian physics, see Garber (1992).

[13] Newton held the view that the Cartesian conception of matter as characterized by expansion just confuses the essential properties of space and bodies with each other. Newton (1726, 1999) considered impenetrability as an important characteristic of the physical bodies and refuted the Cartesian theory with physical arguments. Nevertheless, Koyré (1965) argues for a decisive influence of Descartes on Newton. Even historians of science seem to have their favorite heroes. Newton's struggle with the conceptions of Descartes can be seen perhaps most clearly from the posthumous manuscript, which was probably written before the drafting of the Principia, which is usually quoted by its opening

independent properties or possibilities of action. This made it possible that bodies exert forces without direct spatial contact.[14] Through this, Newton took a step beyond the mechanistic natural philosophy of the seventeenth century which wanted to admit only direct mechanical interactions between bodies.[15] This step was decisive for the further development of physics. This, however, leads to the question, which Newton could not answer, how such a force can be exerted across a spatial distance.[16] This led the deeply religious Newton to a theological pirouette. Whereas the heliocentric model of Copernicus was rejected immediately and vehemently by Martin Luther, and Pope Urban VIII, after much hesitation, finally decided to have Galileo condemnated, Newton's idea that the gravitational effects across spatial distances constitute a proof for the divine control of world affairs was received by enlightened Christians in England as a splendid refutation of the ideas of Descartes that were considered as atheistic. In his physical theory, Descartes had also attempted to explain gravity by vortex motions of contacting and interacting matter particles, that is, by direct physical contact rather than by action at a distance, and thus without any kind of divine agency.[17] But even if the theological twist that Newton had thus given the matter received a lot of support in England at that time, this was of course a scientific impasse.

A central aspect of the following history of ideas and thus a certain guide for our considerations is how this problem of spatial mediation of forces was modified by the electrodynamical concept of a field. This concept replaced the remote with a near action. Via Riemann's new conception of space and its properties, it ultimately led to Einstein's idea of general relativity as a dynamic interaction between space and matter.

The time-dependent positions of these Newtonian mass points could be described by Cartesian coordinates, i.e., by numbers on three mutually perpendicular coordinate axes.

---

words "De gravitatione. . , " and which was first published with English translation in Newton (1962, S. 89–156).

[14] Kepler considered the attraction of the earth as a kind of magnetic force, inspired by the study of magnetism by William Gilbert (1543–1603) and the latter's discovery that the earth also behaves like a magnet, which can explain the properties of the compass. Remarkably, to this day, physics has not succeeded in capturing magnetism and gravitation in a unified theory, as we shall explain in more detail below.

[15] For a succinct but very clear exposition see Westfall (1977). A detailed analysis of the Newtonian concept of force and its historical genesis and preparation can be found in Westfall (1971).

[16] It is a noteworthy fact in the history of science that in the hands of Kepler, this idea was still fertile and pioneering for physics. In particular, this lead to his insight into the cause of the tides. While Galileo had wanted to explain the tides by the earth's rotation and reversely believed to have thereby found a proof for the earth's rotation and thus for the correctness of the Copernican system (although this argument contradicted his own principle of relativity), Kepler attributed the tides to the influence of the moon, that is, to an action over a spatial distance. Had this idea not been accepted, the great system of Newton would not have been possible either. But when the general acceptance of the Newtonian theory made it difficult, if not impossible, to question this idea, the further progress of physics was hindered.

[17] See for instance Koyré (1957). Such views were still advocated by the leading scientist of the eighteenh century, Leonhard Euler, although Euler was deeply religious.

Thus, it was by means of the parametrization of Euclidean space by Cartesian coordinates that a fixed reference system for all physical processes was obtained. For Newton the space—always thought of as Euclidean—thereby also acquired ontological priority over the objects, and Newton then considered space as an attribute of God, as an expression of His omnipresence.[18] Newton also assumed a natural metric for Cartesian space, which thereby (in today's perspective) became Euclidean space. He (or, officially, his spokesperson Samuel Clarke (1675–1729)) criticized Gottfried Wilhelm Leibniz for the latter's purely relational concept of space, arguing that it allowed only qualitative relationships but did not permit quantitative measurements. For Leibniz's geometric ideas, however, it was particularly important not to introduce a scale that was not and could not be conceptually justified. Consequently, a central aim for him was to use only relative proportions, not absolute magnitudes, in the construction of geometry. Thus, Newton's absolute space of Newton was sharply criticized by Leibniz (1646–1716).[19] Leibniz considered spatial relations as relations between objects and came in this way to a relative concept of space, as an order of situations.[20] He could, however, not refute Newton's counter-argument that the rotational motions of liquids demonstrate the physical effect of absolute space (this was not achieved until the nineteenth century when Ernst Mach (1838–1916) explained such physical phenomena by the gravitational effects of the fixed stars[21]).[22] Although Leibniz's ideas contained many seeds for the future development of physics (e.g., the continuity principle and the action

[18] The idea of space as an expression of God's omnipresence had already been developed by the Cambridge Platonists and, in particular, by Newton's friend Henry More (1641–1687) and that of time as an expression of God's eternity and constant presence by Isaac Barrow (1630–1677), the colleague of More and the teacher, colleague and friend of Newton. See the presentation in Burtt (1932). For Newton, space and time even were the sensoria of God, and this naturally led to God's characterization as the self-perception of reality. Here, however, we cannot discuss such later developments in detail, but see for instance my Jost (2019).

[19] See the famous polemics between Leibniz and Newton's follower Samuel Clarke (1675–1729), for example reproduced in Alexander (1977).

[20] For a systematic presentation and analysis of Leibniz's concept of space in the context of his philosophy, I refer to De Risi (2007, S. 283–293). The evolution of Leibniz's thinking about space until its expression in the controversy with Clarke is reconstructed in De Risi (2024). The structural considerations of Leibniz went far beyond the discussion of his time, but because they were not systematically published and therefore not properly understood by his contemporaries, they did not have a sustained effect. For contrasting Leibniz's ideas with contemporary science, see Jost (2019).

[21] But even this argument of Mach did not provide the final resolution. That was supplied only in the general theory of relativity, as will be explained in more detail below.

[22] More generally, Newton argued that one should consider not only kinematic relations, but dynamic effects. Kinematically, one cannot distinguish between a resting a earth and a moving sun, as in the views from Ptolemy to Brahe, or the converse situation, that of Copernicus, where the sun is at rest and the earth moves. But dynamically, the rotation of the earth leads to a flattening at the poles and a buckling at the equator. When Pierre Louis Moreau de Maupertuis (1698–1759) directed a scientific expedition to Lapland to determine the shape of the Earth and verified this Newtonian prediction in 1736, he became instantly famous, and Newtonian physics won the day over the Cartesian view. The physical arguments of Newton for his concept of space are analyzed, for instance, in Fierz (1954).

at proximity or the conservation of energy), Newtonian physics won the day at the time because of its superior force concept. In any case, it was a guiding principle of Newton that the true geometric facts are expressed in the effects of forces. That the sun in the center of the planetary system we recognize from the fact that it holds the planets in their orbits through its gravitational pull. Newton's mathematical formulation of the law of gravitation became the role model of a physical theory, even if the underlying concept of space was problematic and Newton did not feel himself at ease with the concept of action at a distance between remote objects through empty space.[23] The concept of action at a distance was, as already explained, later replaced by the infinitesimal concept of a propagating field in the theories of Faraday (1791–1867) and Maxwell (1831–1879). These theories were concerned with electromagnetism, a physical force different from gravity, but Einstein's general theory of relativity then was a field theory of gravity.

It was certainly a major advance that both Leibniz and Newton, against Aristotle and Descartes, carried out the conceptual separation of space and bodies, decisively, as it seemed at the time. But in a certain sense, this conceptual separation is taken back by the general theory of relativity.

Let me return once more to the dynamical aspects and the role of forces. In the physics of Aristotle and his scholastic followers, a body is moving from its own propensity. A stone is falling to the earth because it has the purpose to get to its natural position. Thus, bodies are actively moving because of intrinsic final causes. The fundamental question then was why when you throw a stone it keeps on moving after it has left your hand that propelled it. For Galileo and Descartes, the question was reversed, that is, why a stone once thrown does not keep on moving forever, but eventually comes to rest. In the physics of Descartes, the physical world is a continuum of contiguous physical bodies, and motion is a mode of a body. In contrast to Aristotle, who did not regard physical motion as fundamentally different from other changes a body can undergo, Descartes restricts the concept of motion to changes of local position, focussing on the relation of a body to its immediate surroundings.[24] The conceptions of both Newton and Leibniz are fundamentally different from those of Aristotle, but in important ways also from those of Descartes. In Newtonian physics, a body changes its state of rest or motion because other, distant, bodies exert forces upon it, and not by its own propensity. What is left is only the Galilean concept of a body's inertia. These

---

[23] Newton did not allow himself in the *Principia* (Newton 1726) to pursue the question of the cause of gravitation. According to his empiricist attitude, through careful observation of the phenomena, he wanted to come inductively to laws which then in mathematical formulation and by mathematical methods allowed for empirically testable prediction of further phenomena. In this sense we should understand his famous quote "Hypotheses non fingo". However, at other occasions, he did speculate about an ether mediating gravity and other physical forces, see Burtt (1932). For instance, on p.350 of the 1979 Dover republication of his *Opticks*, he speculates about an ether that mediates gravitation. The standard version or interpretation of Newtonian physics where gravity acts immediately and across empty space, that is, without the help of a medium, was developed by Newton's followers and derived in a philosophical framework by Kant.

[24] see Garber (1992).

forces could be attractive, like gravity, or repulsive, preventing a body from getting into the space occupied by other bodies. Kant then worked out this dynamical view systematically, replacing concepts like impenetrability of bodies by dynamical actions of such repulsive forces. For him, matter was constituted from a balance between attraction and repulsion.[25] Of course, in Newtonian physics, the interactions between bodies are reciprocal, because of his third law that action equals reaction, but forces are active across spatial distances, as described above. The Leibnizian universe consists of a collection of coexisting monads, each of which reflects the entire universe in itself, although in an imprecise manner. Thus, the world is constituted by a web of interacting monads that feel each others' influences without the need of direct physical contact.[26] In fact, the monads themselves are not spatial, in sharp contrast to Cartesian bodies that are characterized by their spatial extension, but space is only constituted from the web of their interrelations.[27] Thus, for both Newton and Leibniz, a concept of force (from gravity and inertia as empirically observed and quantified as mass times acceleration in Newton's case, intrinsic energy, quantified as mass times the square of velocity and conserved, for Leibniz) is constitutive for matter, in sharp contrast to only geometric extension for Descartes. Bodies are moving (or more precisely, change their state of rest or uniform motion) because they feel the influence of other, distant, bodies. In contrast to Descartes where a body could be influenced only by those other bodies that were in direct physical contact with it, now a body is subjected to the influence of all other bodies. This eventually led the focus from individual bodies to the universe as a whole. This becomes evident, for instance, already in the natural philosophy of Kant.[28] Ultimately, of course, it culminated in Einstein's general theory of relativity. Before that, however, Einstein had first to develop his special theory of relativity where he overcame the assumption of an instantaneous transmission of physical forces or influences across the distance that was underlying Newton's and Leibniz' conceptions. Again, this was prepared by the field theories of electromagnetism, another force that is not transmitted instantaneously, but at a finite speed, that of light.

---

[25] Kant contrasts the mathematical approach of the mechanical philosophy which tries to explain physics through the mathematical analysis of a priori concepts with his own metaphysical-dynamical approach that uses forces as basic ingredients and therefore depends on experience. See Friedman (2012).

[26] Actually, Leibniz' views are more complex, but this is not the place to go into details. See for instance Guerault (1934), and many other texts dealing with Leibniz' natural philosophy.

[27] See for instance Garber (2009), who, however, emphasizes the changes that Leibniz' natural philosophy underwent during his lifetime.

[28] See the comprehensive analysis of Kant's natural philosophy by Friedman (2012).

2.2 Kant's Philosophy of Space

Newtonian physics and Leibniz's ontology then were the starting point for Immanuel Kant.

Kant wanted to develop a philosophical justification of Newtonian physics. He dissolved the contrast between Leibniz's pure relational understanding of space and Newton's absolute space by not locating space in the material world, but as a form of intuition into the perceiving subject.[29] In any case, Kant emphasized the relativity of space against Newton. But Kant's main point is that space as the possibility of coexistence is a precondition of experience. For Kant, space in this sense is empirically real, but transcendentally ideal because it does not constitute a basis for the things by themselves. Kant then goes one major step further and considers statements about space as synthetic judgments a priori, that is, constructions of the perceiving subject that lie before every experience (and vice versa make experience as such possible). That they are synthetic means that they cannot be simply obtained through an analysis of the concept of space, but rather are autonomous constructions. For Kant, these synthetic judgments a priori include the axioms of Euclidean geometry. Against Leibniz and Wolff (1679–1754, a very influential philosopher in the first half of the eighteenth century who built upon the ideas of Leibniz), Kant thus emphasizes and elaborates the axiomatic nature of geometry, i.e., that geometry has real axioms[30] and that the propositions of geometry cannot be obtained analytically from definitions. For this essential insight that was also accepted by mathematics, the contacts of Kant with the mathematician Johann Heinrich Lambert (1728–1777), a precursor of non-Euclidean geometry, were probably also helpful. In particular, Euclidean geometry, for Kant, is not logically necessary.

Moreover, Kant emphasizes the constructive nature of geometry and derives from that the uniqueness of three-dimensional Euclidean geometry as intuitively constructible. Whether Euclidean geometry thus in Kant's view is necessary for our thinking, or a necessary precondition for our experience, is a much discussed issue that is of central importance for the interpretation of Kant. After all, Riemann implicitly points out that the assumptions of Euclidean geometry are not necessary, but represent specific geometric hypotheses, and Helmholtz makes this point the heart of his epistemological argument. The orthodox Kantians therefore initially rejected the ideas of Riemann and Helmholtz.[31] But when the unten-

---

[29] Immanuel Kant, edition of Weischedel (Kant 1977). I shall use the translation (Kant 1998) This edition, like that of Weischedel, gives the paginations from both the first (from 1781) and the second edition (from 1787) of the *Critique*; for instance, A86/B118 means p.86 of the first and p.118 of the second edition.

[30] Where axioms here should not be interpreted in a modern sense, following Hilbert, as arbitrary stipulations.

[31] In this context, it can only contribute to confusion when Paul Franks in Leiter and Rosen (2007, pp. 243–286) (about Helmholtz especially pp. 269–276) classified Helmholtz as a Neo-Kantian, because the so-called Neo-Kantians were his most important philosophical adversaries, besides people like Hering whom he referred to as a nativist. In Schiemann (1997), Schiemann works out especially how Helmholtz' approach of a justification of experience in the conditions of physical measurements differs from the Kantian point of departure of the transcendental subject, and examines the systematic

ability of this position gradually became clearer, efforts were made later to incorporate the arguments of Riemann and Helmholtz into the Kantian system.[32]

Because this is an important aspect of the reception history, it is necessary to represent the view of Kant in more detail. More precisely, this is the theory of space developed in the *Critique of Pure Reason*; we should note that Kant's concept of space has changed several times in the course of his life, always wrestling with the relationship between Newtonian physics with its absolute concept of space and Leibniz's ontology in which space was only a relation, and therefore not real, but rather ideal. This is all the more complicated because both Newton and Leibniz mix this discussion with theological aspects. In his early work, *Gedanken von der wahren Schätzung der lebendigen Kräfte* (*Thoughts on the True Estimation of Living Forces*) Kant poses the hypothesis of a relationship between the forces operating in space and its geometric structure, and in particular between the law of gravitation and the three-dimensionality of the space. He also contemplates the possibility of higher dimensional spaces.[33]

But in his dissertation, Kant then argued for the ontological priority of space over the objects contained in it.[34] He employs the example of the left and right hands (or one hand and its mirror-image or a left and a right glove, or a left-handed and a right-handed screw), which by themselves are similar—in modern mathematical terminology isomorphic to each other –, but nevertheless different from each other insofar as they cannot be made to coincide in space. This means, according to Kant, that their properties are not completely determined by themselves, but rather that there is an important property, handedness, that is only assigned to them by space. The last and for Kant's purposes important part of the argument cannot be upheld, however. To see this, a deeper insight into the structure of space is required that was not yet available to Kant. To understand this, consider a dimensionally reduced version of a left and a right handprint in the Euclidean plane. These figures cannot be converted into each other by a movement in the plane (or in a different interpretation, by a movement of the plane). But this is not a property of the two figures, but depends on the topological structure of the space. If we glue a planar strip, contain our two figures, into a Möbius strip, it becomes possible in this new geometric space to transform the two figures into each other. The difference between the plane and the Möbius strip which both, as will be explained below, have the same internal geometry, because the figures don't get distorted by the construction of the Möbius strip in any way, is that the latter is not orientable. This means that handedness can no longer be assigned in a consistent manner. Therefore the geometrical difference between the two figures disappears. Another possibility to transform

changes that Helmholtz' natural philosophy 'underwent in the course of his life. See also some essays in Cahan (1993).

[32] See the references below in the reception history.

[33] This had already been contemplated by Leibniz, see De Risi (2007). Leibniz then went on to attempt to prove the three-dimensionality of space.

[34] Kant (1977a), S. 991–1000; English translation in Kant (1992). This example is taken up again in Kant (1977b), pp. 111–264, §13; English translation in Kant (2004).

one of the figures into the other results if we move out of the plane into the surrounding space and flip the figures over. Geometrically, this is a reflection of the plane across a straight line, an operation that cannot be realized as a continuous motion in the plane itself, but only as a transformation in three-dimensional space. So if we either deprive space of its orientation, that is, make it nonorientable, or add a dimension, we can move the left into the right figure, and left and right handedness cease to be properties of the figures. The same is possible in three-dimensional space. Like the Möbius strip in two dimensions, we can also mathematically define a three-dimensional non-orientable space with local Euclidean geometry, and we can also make the transition from three-dimensional to four-dimensional space, to be able to convert a left into a right hand by a movement in space. Handedness therefore is not an absolute property of geometric objects that is assigned to them by (the) space, but the possibility of a distinction between left and right handedness is a topological property of space.[35] In Kant's example, this property of space is detected by observations on the objects found in space. This precisely questions the ontological priority of space.[36] This issue, however, could be clarified only through the geometrical insights of Gauss (1777–1855) and Riemann. Gauss[37] argued already against Kant that Kant's own remarks that we can communicate our realization of this difference between left and right to others only by its demonstration at really existing objects just proves that space must have a real meaning which is independent of our intuitions. This argument is not (or not only) directed against the ontological priority of space over objects, but against the doctrine developed in the *Critique of Pure Reason* of space as pure intuition of external experience which represents a significant change compared to Kant's dissertation. In fact, Kant famously argued that space is a necessary intuition a priori which is a precondition for all external experience, because one can indeed imagine a space without objects, but not that there is no space. Space is an intuition, not a concept, because we can reach non-obvious conclusions about it, the geometrical propositions.[38] It is furthermore a pure, not an empirical intuition, because the geometrical propositions are apodictically true, i.e., associated with the insight into their necessity like, in particular, the three-dimensionality of space. An example, to which we shall need to return below is "That the straight line between two points is the shortest is a synthetic proposition. For my concept of *the straight* contains nothing of quantity, but only

35 see for example Weyl (2009).

36 For a comparison of the positions of Leibniz and Kant on this issue and a newer overview of the literature on the subject, refer to De Risi ([pp. 283–293]2007).

37 Gauss ([Bd. II, p. 177]1973).

38 The fact that from mathematical axioms conclusions can be drawn that are not obvious, is a central theme of the philosophy of mathematics. The Platonic approaches view mathematics as an opportunity or a tool to discover eternal truths. Weyl (2009), however, emphasizes the constructive and creative nature of mathematics.

a quality. The concept of the shortest is therefore entirely additional to it, and cannot be extracted out of the concept of the straight line by any analysis. Help must here be gotten from intuition, by means of which alone the synthesis is possible."[39]

In particular, the mathematical intuition about space is not empirical: "it is not images of objects but schemata that ground our pure sensible concepts. No image of a triangle would ever be adequate to the concept of it. For it would not attain the generality of the concept, which makes this valid for all triangles, right or acute, etc., but would always be limited to one part of this sphere. The schema of the triangle can never exist anywhere except in thought, and signifies a rule of the synthesis of the imagination with regard to pure shapes in space."[40] But since this intuition is not empirical, it must come from the perceiving (or more precisely, according to Kant, the transcendental) subject. The necessity of geometrical propositions thus has its origin in the perceiving subject, as a condition for the possibility to organize the multitude of phenomena in their spatial distinctiveness. In that regard and to that extent, the propositions of Euclidean geometry are not logically necessary. For Kant, Euclidean geometry is only distinguished by the fact that it is intuitively constructible. We humans necessarily imagine space to be Euclidean. Kant utilizes the following example to bring this point home: "Thus in the concept of a figure that is enclosed between two straight lines there is no contradiction, for the concepts of two straight lines and their intersection contain no negation of a figure; rather the impossibility rests not on the concept in itself, but on its construction in space, i.e., on the conditions of space and its determinations; but these in turn have their objective reality, i.e., they pertain to possible things, because they contain in themselves a priori the form of experience in general."[41]

However, the interpretation of Kant was quite fluid with regard to these points. One reason, of course, is that Kant himself had changed his views on these matters several times and the crucial arguments in the *Critique of Pure Reason* use terms that become clear only from his later reasonings. On the other hand, it caused considerable difficulties for the Kantians to develop an interpretation of relevant arguments of Kant that did not conflict with subsequent mathematical and physical insights and findings.

2.3 Euclidean Space as the Basic Model

We now want to place some of the developments described above into the systematic context opened up by Riemann's work. Thus, in this section, we no longer proceed historically, but rather in terms of concepts. In the geometry of Riemann and his followers and successors, on the one hand, the priority of Euclidean space is abandoned, but on the other hand, Euclidean space continues to enjoy a special role as a reference model. As shall be explained in

[39] Kant ([2nd ed., Introduction, B16 (p.145)]1998) (emphasis in the original).

[40] Kant (1998, A141/B180 (p. 273)).

[41] Ibid., A221/B268 (p.323).

Chap. 4 and mathematically formalized in Sect 4.4, a Riemannian geometry is characterized by the fact that it is infinitesimally, that is, regarding the infinitely small, Euclidean. (It is, however, no longer necessarily locally Euclidean, because of the possibility of curvature, nor globally, that is, on the large scale, because of the possibility of other topological properties and types.) Curvature measures the local deviation from the Euclidean model. Curvature is thus normalized in such a manner that the Euclidean space has curvature equal to zero.[42] After this anticipation of a basic concept of Riemannian geometry, we want to once again return to the question of how the Euclidean space historically could gain its role of a null model. Here again, different strands of development merge.

1. We have already described the development of the linear, that is, Euclidean, perspective in the theory and practice of Renaissance painting which in turn rested on the Euclidean laws for the propagation of light rays. In the underlying idea of the projection of Euclidean space onto a plane, parallel straight lines are conceived of as intersecting at infinity, and such an intersection point that is infinitely far away, i.e., that represents a pencil of parallel straight lines, is contracted into the vanishing point.
2. We have also explained how the notion of Euclidean space as a carrier of physical processes had emerged from gravitational theory. We want to pick this up again because it is also essential for a deeper understanding of the basis of the subsequent mathematical development. The law of inertia states that a body on which no external forces act moves without acceleration and therefore in particular straight in a space that is mathematically conceived as Euclidean and physically imagined as empty. It is now important to realize that such a situation is actually unphysical because physical processes by their very nature are interactions between bodies. Thus, Galileo had refused to use such a situation as the basis of his physical theory. In his models he places (in a later conceptualization, because Galileo of course did not possess a theory of gravity) an otherwise force-free movement into a central gravitational field. So he was not ready to accept a body on which no gravitational forces act as the most basic situation, because such a body does not possess any physical reality.[43] So in his reasonings, he often replaced the infinite (Euclidean) plane by a spherical surface on which a body that is solely exposed to a central gravitational force moves otherwise freely. A crucial step which Galileo ultimately was not ready to take, thus was to conceive a physical situation as a deviation from an unphysical null model. Newton, in contrast, had, as described, attributed an ontological reality as absolute space to just this null model.

[42] Euclidean space is also referred to as "flat", and the word "curvature" should then simply express the deviation from this plane, straight shape.

[43] Koyré (1966) therefore tries to deny Galileo knowledge of the law of inertia, even if this law is implicitly assumed and explicitly expressed in the passages quoted by him from Galileo and his successors Cavalieri (1598–1647) and Torricelli (1608–1647) and Gassendi (1592–1655) several times. Simply, unlike Newton, he did not make this the basis of his physical theory, because he had considered a motion without the influence of other bodies as unphysical.

3. Since the Euclidean space can be thought of as physically empty, in physical terms it then is the vacuum, or mathematically, the substrate of the vacuum. The question of the possibility of the vacuum now also touches on the basis of physical theory, and, as outlined, for example, both Aristotle and Descartes rejected the vacuum, because it is not compatible with their physical theories. Descartes failed as a physicist in particular because his mathematical concepts did not fit together with his physical notions. His great mathematical achievement was the introduction of the Cartesian coordinate space for the systematic description and representation of algebraic equations.[44] This space then made possible the systematic treatment of functional relationships by Cartesian graphs. The three-dimensional Cartesian space is a Euclidean space in which the position of a point is determined by numerical values on three mutually perpendicular coordinate axes.[45] The Cartesian space therefore is actually ideally suited for the description of the vacuum, at least if it properly captures the topological and dimensional properties of the vacuum, as was implicitly assumed at that time.[46] The physics of Descartes, however, was based on mechanical interactions due to collisions. Physics was therefore not possible for him in a vacuum. Galileo, however, is in the tradition of atomism, which had been proposed and developed in antiquity by Leucippus (fifth century BC), Democritus (ca. 460–370 BC) and Epicurus (341–270 BC) (but rather as a natural-philosophical speculation without any concrete physical basis), and for him, the vacuum is therefore unproblematic. The Newtonian theory then, as explained, raised the question as to what extent empty space can be the carrier of physical forces.
4. The Euclidean space is not only empty, but also unbounded and infinite. The difference between these two properties again was not clear before the work of Riemann, who clarified that manifolds need not have a boundary without necessarily having to be infinite (in modern terminology, this would be compact closed manifolds such as the surface of the sphere and their higher-dimensional analogues, see the explanatory notes to Riemann's text below). The infiniteness of space had also been rejected for a long time from natural philosophical and theological considerations, from Aristotle to Kepler. The

[44] The Cartesian coordinates, however, are only implicit in the geometry of Descartes and were not constructed by him explicitly. But since Descartes laid the conceptual foundations, it is still justified to name these coordinates after him. See, for example Giaquinta (2010) or Ostermann and Wanner (2012).

[45] As we shall explain below, the actual logical relationship is rather the other way around: One obtains the metrical structure of Euclidean space by interpreting the magnitudes of coordinate differences on each coordinate axis of a Cartesian space as distances and by declaring different coordinate axes as perpendicular to each other. Thus, Euclidean space possesses a metric structure which as such is not yet contained in the Cartesian concept, while the Cartesian coordinate space is determined in a way not provided by the Euclidean concept. The clear separation of geometric facts and their different descriptions in different coordinate systems is then just one of the essential achievements of Riemann. See also Jost (2015, pp.123–127) on this point.

[46] The question of whether it is appropriate to attribute to the vacuum the geometric structure of Euclidean space leads into modern physics, which will be discussed below.

idea of an infinite space was prepared by Nicholas of Cusa (1401–1464) and emphatically embraced by Giordano Bruno as a liberation from the limitations of the medieval world view.[47]. It is then remarkable that such an infinite space could become the reference model for finite spaces (compact Riemannian manifolds).

5. The infinitesimal calculus introduced by Leibniz and Newton can be considered and utilized as a linear approximation scheme for possibly nonlinear processes. A process is thus infinitesimally linear in this scheme, and the linear structure at a given time t is determined by its derivative with respect to t. Locally, however, the process deviates from this linear approximation due to interactions. Differential calculus, which thus had originally been developed for analyzing temporal processes, then became a general tool for approximating also static structures, in particular in the hands of Leonhard Euler (1707–1783), the mathematician who dominated the eighteenth century. Differential geometry and in particular Riemannian geometry will then model a general spatial structure as infinitesimally linear and quantify the local deviation in the vicinity of a given point p from its linear approximation by the curvature of that space at p. The Euclidean-Cartesian space is then distinguished in that it globally and not only infinitesimally carries a linear structure (it is a vector space in modern mathematical terminology). Thus again, Euclidean space becomes the model space with which a general space can be compared.[48] Furthermore, Riemann will transfer the global coordinatisation of Cartesian space into the local coordinate description of a manifold (a Riemannian concept illustrated below). Coordinates thus become, rather than an ontological basis, a conventional description of geometrical relations and physical processes. This will then in turn lead to the fundamental question of Einstein's theory of relativity, to identify the geometrical and physical properties that are independent of the choice of coordinates. The rules for the transformation between different coordinate descriptions, which were in particular systematically developed by the successors of Riemann, and Riemann's idea of curvature invariants provide the mathematical foundation for Einstein's theory.
6. The space-time underlying Einstein's theory of special relativity is Minkowski space-time, a version of four-dimensional Euclidean space, but with a relativistic metric where time and space carry opposite signs, but which retains the vector space structure of Euclidean space, and which contains three-dimensional Euclidean space as a subspace. In Einstein's theory of general relativity, Minkowski space-time still plays a fundamental

---

[47] On this, see Koyré (1957). It is remarkable that cosmology today returns to the idea of a finite cosmos, amongst other reasons, to "explain" the emergence of the universe from a singular beginning, the Big Bang, and thus to regain the historical dimension in contrast to a truely infinite, but static universe.

[48] However, more general spatial concepts were then introduced after Riemann, which give up this condition of the approximability by a Euclidean space. An example are the so-called topological spaces. Also the concept of a Riemannian manifold is later developed to the extent that only so-called differentiable manifolds, but no longer more general manifolds satisfy this approximability condition. Thus, Euclidean space will finally lose its special role. More details will be presented below, when we analyze Riemann's text.

role as a reference space. In particular, it is a vacuum solution of the Einstein field equations. It is not the only vacuum solution, but it is nevertheless the simplest and most basic such solution.

7. The Hilbert space of quantum mechanics is an infinite-dimensional Euclidean space. In particular, it carries a Euclidean metric structure.

After this somewhat lengthy anticipation, we shall now return to the historical development before Riemann.

2.4 The Development of Geometry: Non-Euclidian and Differential Geometry

A guiding problem in the development of geometry was the parallel problem. The 5th postulate or 11th axiom of Euclid states that "If a line segment intersects two straight lines forming two interior angles on the same side that sum to less than two right angles, then the two lines, if extended indefinitely, meet on that side on which the angles sum to less than two right angles".[49]

When one assumes the other Euclidean axioms, an equivalent formulation of the Euclidean parallel postulate, sometimes named after the Scottish mathematician John Playfair (1748–1819), is: "In a plane, given a line and a point not on it, at most one line parallel to the given line can be drawn through the point". This is in turn equivalent to that the sum of the angles is exactly 180 degrees in each triangle. The parallel postulate obviously has a special position within Euclid's work, and therefore the question arose whether it can be derived from the other axioms and postulates, and thus would not be independent from these. After intense but ultimately unsuccessful attempts to deduce a contradiction from the assumption that this axiom does not hold, and thus to demonstrate its dependence on the others, it slowly dawned that an alternative to Euclidean geometry is logically possible, in which the parallel axiom is not valid. After important precursors like Giovanni Girolamo Saccheri (1667–1733) (Saccheri 2014) and Johann Heinrich Lambert (1728–1777), the first person who fully realized this was Carl Friedrich Gauss, who, however, for fear of being misunderstood by his contemporaries, did not want to make his findings public, and then non-Euclidean geometry was independently discovered by Nikolai I. Lobatchevsky (1792–1856) and Janos Bolyai (1802–1860) in the years before 1830.[50] The founders of non-Euclidean geometry came to realize that which geometry is valid, a Euclidean or a non-Euclidean one,

[49] The standard English reference is the translation of Heath of Heiberg's translation from Old Greek (Heath 2000).

[50] English translations in Bonola (1955). For more information, refer to the bibliography.

is an empirical question that could be decided by angle measurements in triangles in space. Even on an astronomical scale, however, with the measurement accuracy available at the time, no deviation from the Euclidean angle sum could be found.[51]

However, the geometric starting point of Riemann was not the non-Euclidean geometry, of which Riemann apparently had not even taken note,[52] but rather the theory of surfaces developed by Carl Friedrich Gauss.[53]

When conducting a geodesic survey of the Kingdom of Hanover, Gauss wanted to penetrate this problem from a theoretical perspective and therefore examined the geometry of surfaces in Euclidean space. Of far-reaching importance was his distinction between geometric quantities that can be determined already by measurements on the surface itself, and those for whose determination measurements outside the surface in the surrounding space are needed. This is the distinction between the internal and the external geometry of surfaces in space. As a fundamental quantity of internal geometry Gauss identified that quantity which was later named Gaussian curvature or Gauss curvature. For this quantity, Riemann then gave a novel interpretation and a far-reaching generalization. The initial starting point for Gauss were quantities of the external geometry, the so-called principal curvatures of a surface S at a given point P, originally introduced by Euler (1707–1783). For their determination, one considers the planes that perpendicularly intersect S at P. The intersection between such a plane and S (also called a normal section) is then (near P) a curve c on S. This curve then has a curvature k (measured with a sign, that is, it can be positive or negative) at P. Among all these curves of intersection, there is a smallest curvature k_1 and a largest curvature k_2 .[54] These two principal curvatures in general depend on the shape of the surface in space. Gauss then derived the remarkable result (which he called the *Theorema egregium*) that the product $K = k_1 \cdot k_2$ no longer depends on how the surface sits in space. Thus, it is a quantity of the internal geometry. In particular, the Gaussian curvature is thus invariant under bending the surface, as long as one does not stretch or compress it. For example, one can roll a piece of paper into a cylinder or a cone-shaped container, and this does not change the Gaussian curvature, which in this case is and remains 0. In contrast, a spherical surface has positive Gaussian curvature K, and the curvature is the greater, the smaller the radius of the sphere (K is inversely proportional to the square of the radius). Because the Gaussian curvature is bending invariant it follows from the different values of K for the plane and the sphere that a flat surface cannot be brought into a spherical shape without

[51] For details, see for instance Torretti (2012, pp. 63f, 381).

[52] See Scholz (1982).

[53] Gauß (1828). Reprinted in Gauss (1973, Bd. 4, 217–258); English translation in Dombrowski (1979).

[54] This was first shown by Euler (1955, pp. 1–22). Unless all the normal sections have the same curvature, these two intersection curves of extremal curvature are uniquely determined and intersect each other at right angles.

stretching. A saddle-shaped surface has negative Gaussian curvature, because in this case, the two principal curvatures are of opposite sign, since the two perpendicular intersections with planes curve in opposite directions.[55]

Gauss then also established a relationship between the sum of the angles in a triangle formed by shortest lines on a surface and the integral of K over this triangle (*Theorema elegantissimum*). Here, then, there is a direct relation with the non-Euclidean plane. This plane is nothing but the intrinsic geometry of a surface of constant negative curvature, and the sum of the angles in a triangle is therefore less than 180 degrees. Gauss himself has probably already seen this connection, but its real significance became clear only by the work of Riemann (although Riemann had not even taken notice of non-Euclidean geometry).

2.5 The Story of Riemann's Habilitation Address

While Riemann, in addition to his proper mathematical research, had immersed himself also into natural philosophical speculations and thereby, in many respects, had considered mathematics, physics and natural philosophy as a whole,[56] the fact that the text presented here saw the light nevertheless came about rather by a coincidence. As is still common today in German universities, Riemann had to submit three different themes for his habilitation address that the faculty could choose from. But usually the first topic was selected. Riemann then chose the first two topics from his current mathematical research and then named as

[55] For a modern presentation, see for instance Eschenburg and Jost (2013).

[56] In his private notes, Riemann cited in particular the philosopher Johann Friedrich Herbart (1776–1841) and mentions him also at the beginning of his habilitation address, see Herbart (1964), in particular *Psychologie als Wissenschaft*, 2 parts, Vol. 5, 177–402, and Vol. 6, 1–339 (originally published in 1824/25). In 1809 Herbart became Kant's successor as the chair for philosophy in Königsberg and in 1834 took over the chair of philosophy in Göttingen. He represents the transition from idealism to realism in German philosophy of the nineteenth century. He criticizes Kant from an empiricist and association psychology position. The individual being is for him a unit, a feature bundle that by coming together with others acquires different characteristics that in each case can represent different continua. Thus snow is white, when the eye sees it, cold, when the hand touches it. These continua can be conceived spatially. He emphasizes in particular the historical contingency and conditionality of the notion of space, which for him, according to his considerations just presented, was only an example of a continuous sequel. The relationship between the ideas of Herbart and the concepts Riemann is discussed in Erdmann (1877, S.29–33) and Boi (1995, S.129–136). Scholz (1982) on the other hand comes to the conclusion that ultimately the influence of Herbart's thoughts on the Riemannian manifold concept was rather minor, even if Riemann may have been guided by some general principles of Herbart, as that for each area of science a main concept needs to be worked out, or that ideas such as tone or color are not only quantitatively different, but also subject to different types of mathematical laws and consequently should be investigated by the methods of mathematics. In this context, we also refer to the presentation in Pulte (2005, S. 375–388).

the third theme the one of the foundations of geometry. To his surprise, at the instigation of Gauss,[57] the faculty chose the last topic, and to prepare the corresponding lecture within the prescribed period caused Riemann considerable effort. The lecture was held on June 10, 1854. Gauss, otherwise very difficult to impress, was extremely impressed by Riemann's lecture. Nevertheless, Riemann could not bring himself to publish this lecture; this was only done posthumously in 1868 by Richard Dedekind.

References

Alexander, Robert Gavin. 1977. *The Leibniz-Clarke Correspondence*. Manchester University Press.

Algra, Keimpe. 1994. *Concepts of Space in Greek Thought*. Brill.

Andersen., Kirsti. 2008. *The Geometry of an Art: The History of the Mathematical Theory of perspective from Alberti to Monge*. Berlin: Springer.

Boi, Luciano. 1995. *Le problème mathématique de l'espace: une quête de l'intelligible*. Springer.

Bonola, Roberto. 1955. *Non-Euclidean Geometry: A Critical and Historical Study of Its Development*. Dover.

Burtt, Edwin Arthur. 2003. *The Metaphysical Foundations of Modern Science*. Dover (Reprint of the 2n ed. of 1932).

Cahan, David, ed. 1993. *Hermann von Helmholtz and the Foundations of Nineteenth-Century Science*. University of California Press.

Cassirer, Ernst. 1957. *Das Erkenntnisproblem in der Philosophie und Wissenschaft der neueren Zeit*. 4 Vols. Wiss. Buchgesellschaft, Darmstadt, 1974 (Reprint of 3rd ed. of Vols. 1,2 from 1922, 2nd ed. of Vol. 3 from 1923, 2n ed. of Vol. 4 from 1957).

Cohn, J. 1896. *Geschichte des Unendlichkeitsproblems im abendländischen Denken bis Kant*. Leipzig: Wilhelm Engelmann.

De Risi, Vincenzo, ed. 2015. *Mathematizing Space*. The Objects of Geometry from Antiquity to the Early Modern Age: Birkhäuser.

De Risi, Vincenzo. 2007. *Geometry and Monadology: Leibniz's Analysis Situs and Philosophy of Space*. Birkhäuser.

De Risi, Vincenzo. 2016. Francesco Patrizi and the new geometry of space. In *Boundaries, Extents and Circulations: Space and Spatiality in Early Modern Natural Philosophy*, 55–106. Springer, Berlin.

De Risi, Vincenzo. 2024. The genesis of relationism. Leibniz's early theory of space and Newton's scholium. *Oxford Studies in Early Modern Philosophy*, 12: 98–143.

Dombrowski, Peter. *150 years after Gauß' Disquisitiones generales circa superficies curvas*, volume 62 of *Astérisque*. Soc. Math. France.

Edgerton, Samuel Y . 2009. *The Mirror, The Window, and the Telescope: How Renaissance Linear Perspective Changed Our Vision of the Universe*. Cornell University Press.

Eisler, Rudolf. 1912. *Philosophenlexikon*. Berlin: Mittler.

Erdmann, Benno. 1877. *Die Axiome der Geometrie. Eine philosophische Untersuchung der Riemann-Helmholtzschen Raumtheorie*. Leopold Voss, Leipzig.

Eschenburg, Jost-Hinrich, and Jürgen Jost. 2013. *Differentialgeometrie und Minimalflächen*, 3rd ed. Springer.

Euler, Leonhard. 1955. *Opera omnia*, vol. XXVIII. 1st series. Lausanne: Orell Füssli.

[57] See the corresponding quote from the Dean's Office Archive in Laugwitz (1996, p. 218).

Field, Judith Veronica. 1997. *The Invention of Infinity: Mathematics and Art in the Renaissance*. New York: Oxford University Press.
Fierz, Markus. 1954. Über den Ursprung und die Bedeutung der Lehre Isaac Newtons vom absoluten Raum. *Gesnerus* 11 (3–4): 62–120.
Friedman, Michael. 2012. *Kant's Construction of Nature: A Reading of the Metaphysical Foundations of Natural Science*, 2nd ed. Cambridge University Press.
Garber, Daniel. 1992. *Descartes' Metaphysical Physics*. University of Chicago Press.
Garber, Daniel. 2006. Physics and foundations. In *The Cambridge History of Science: Early Modern Science*, ed. Lorraine Daston, and Katherine Park, vol. 3, 21–69. Cambridge University Press.
Garber, Daniel. 2009. *Leibniz: Body, Substance, Monad*. Oxford University Press.
Gauss, Carl Friedrich. 1973. *Werke*. Dieterichsche Universitätsdruckerei, Göttingen, 1870–1927 (Reprint Olms, Hildesheim, New York, 1973).
Gauß, Carl Friedrich. 1828. Disquisitiones generales circa superficies curvas. *Commentationes Societatis Gottingensis*, 99–146.
Giaquinta, Mariano. 2010. *La forma delle cose*. Ed. di storia e letteratura, Roma.
Guerault, Martial. 1934. *Dynamique et métaphysique leibniziennes*. Paris: Les Belles Lettres.
Heath, Thomas L. 2000. *The Thirteen Books of Euclid's Elements*, trans. from the Text of Heiberg, Dover.
Herbart, Johann Friedrich. 1964. *Sämtliche Werke*. 19 Bde., in chronologischer Reihenfolge herausgegeben von Karl Kehrbach und Otto Flügel, Langensalza, 1882–1912; reprinted Aalen, Scientia-Verlag.
Jost, Jürgen. 2015. *Mathematical Concepts*. Berlin: Springer.
Jost, Jürgen. 2019. *Leibniz und die moderne Naturwissenschaft*. Berlin: Springer.
Kant, Immanuel. 1977. *Kritik der reinen Vernunft (1781)*, in: *Werkausgabe Bd. III/IV*. (Ed. Wilhelm Weischedel) Suhrkamp, Frankfurt/M.
Kant, Immanuel. 1977. *Prolegomena zu einer jeden künftigen Metaphysik, die als Wissenschaft wird auftreten können (1783)*, in: *Werkausgabe Bd. V*. (Ed. Wilhelm Weischedel) Suhrkamp, Frankfurt/M.
Kant, Immanuel. 1977. *Von dem ersten Grunde der Unterschiede der Gegenden im Raume (1768)*, in: Werkausgabe Bd. II. (Ed. Wilhelm Weischedel) Suhrkamp, Frankfurt/M.
Kant, Immanuel. 1992. *Theoretical Philosophy, 1755-1770*. Transl. and ed. D. Walford, with R. Meerbote. Cambridge University Press.
Kant, Immanuel. 1998. *Critique of Pure Reason*, transl. by by Paul Guyer and Allen W. Wood. Cambridge University Press.
Kant, Immanuel. 2004. *Prolegomena to Any Future Metaphysics That Will Be Able to Come Forward as Science*. Transl. and ed. G. Hatfield. Cambridge University Press (rev. ed.).
Koyré, Alexandre. 1965. *Newtonian Studies*. Harvard University Press.
Koyré, Alexandre. 1937. *Galilée et l'expérience de Pise: á propos d'une légende*. Sorbonne: Annales de l'Université de Paris.
Koyré, Alexandre. 1957. *From the Closed World to the Infinite Universe*. Baltimore: Johns Hopkins University Press.
Koyré, Alexandre. 1966. *Études galiléennes*. Paris: Hermann.
Laugwitz, Detlef. 1996. *Bernhard Riemann 1826–1866*. Basel: Birkhäuser.
Leiter, Brian and Michael Rosen, ed. 2007. *The Oxford Handbook of Continental Philosophy*. Oxford University Press.
Newton, Isaac. 1726. *Philosophiæ naturalis principia mathematica*, 3rd ed . London: Innys.
Newton, Isaac. 1962. *Unpublished Scientific Papers of Isaac Newton: A Selection from the Portsmouth Collection in the University Library, Cambridge*, ed. A. R. Hall and M. Boas Hall. Cambridge University Press.

Newton, Isaac. 1999. *The Principia: Mathematical Principles of Natural Philosophy*, transl. B. Cohen and A. Whitman. University of California Press.

Ostermann, Alexander, and Gerhard Wanner. 2012. *Geometry by Its History*. Springer.

Panofsky, Erwin. 1991. *Perspective as Symbolic Form*. transl. Christopher Wood. Zone Books.

Panofsky, Erwin. 1927. *Perspektive als "symbolische Form"*. Leipzig/Berlin: Teubner.

Pulte, Helmut. 2005. *Axiomatik und Empirie. Eine wissenschaftstheoriegeschichtliche Untersuchung zur Mathematischen Naturphilosophie von Newton bis Neumann*. Wiss. Buchges., Darmstadt.

Redondi, Pietro. 1987. *Galileo Heretic*. Princeton University Press.

Redondi, Pietro. 1983. *Galileo eretico*. Torino: Einaudi.

Saccheri, Gerolamo. 2014. *Euclid Vindicated from Every Blemish*. Classic Texts in the Sciences. Edited and Annotated by Vincenzo De Risi. Translated by GB Halsted and L. Allegri. Springer

Schiemann, Gregor. 1997. *Wahrheitsgewissheitsverlust. Hermann von Helmholtz' Mechanismus im Anbruch der Moderne. Eine Studie zum Übergang von klassischer zu moderner Naturphilosophie*. Wiss. Buchgesellschaft, Darmstadt.

Scholz, Erhard. 1982. Riemanns frühe Notizen zum Mannigfaltigkeitsbegriff und zu den Grundlagen der Geometrie. *Archive for History of Exact Sciences*, 213–232.

Scholz, Erhard. 1982. Herbart's influence on Bernhard Riemann. *Historia mathematica* 9 (4): 413–440.

Torretti, Roberto. 2012. *Philosophy of Geometry from Riemann to Poincaré*. Springer.

Westfall, Richard S. 1971. *Force in Newton's Physics: The Science of Dynamics in the Seventeenth century*. London: MacDonald.

Westfall, Richard S. 1977. *The Construction of Modern Science: Mechanisms and Mechanics*. Cambridge University Press.

Weyl, Hermann. 2009. *Philosophy of Mathematics and Natural Science*, trans. from the German. Princeton University Press.

3 Riemann's Text

I am aware of three English translations of Riemann's habilitation address. The first one was presented by the eminent British geometer William Kingdon Clifford (1845–1879) in *Nature* (Riemann 1873), and is reproduced in his *Mathematical papers* (Clifford 1882).

The second translation was carried out by Henry S. White in Riemann (1959). In this translation, the only formula from Riemann's text, that which gives the constant curvature metrics, is badly garbled. Therefore, I have not chosen this one for the present volume.

Finally, there is a more recent translation with a detailed commentary by the differential geometer Michael Spivak in Vol. 2 of his *A comprehensive introduction to differential geometry* (Spivak 1970–1975). Spivak's textbook is readily available in mathematical libraries.

The English of Clifford may appear somewhat old-fashioned for a modern reader. For instance, he writes "manifoldness" instead of the simpler modern translation "manifold" of Riemann's term "Mannigfaltigkeit". But Riemann's German sounds likewise somewhat old-fashioned, and for that matter, "manifoldness" is the more accurate translation of Riemann's term. In any case, for historical reasons, I have selected that translation here.

© The Author(s), under exclusive license to Springer Nature Switzerland AG 2025

J. Jost, *Bernhard Riemann — On the Hypotheses Which Lie at the Bases of Geometry*, Classic Texts in the Sciences, https://doi.org/10.1007/978-3-031-91489-8_3

ON THE HYPOTHESES WHICH LIE AT THE BASES OF GEOMETRY

(Translated by William Kingdon Clifford)

Plan of the Investigation

It is known that geometry assumes, as things given, both the notion of space and the first principles of constructions in space. She gives definitions of them which are merely nominal, while the true determinations appear in the form of axioms. The relation of these assumptions remains consequently in darkness; we neither perceive whether and how far their connection is necessary, nor, a priori, whether it is possible.

From Euclid to Legendre (to name the most famous of modern reforming geometers) this darkness was cleared up neither by mathematicians nor by such philosophers as concerned themselves with it. The reason of this is doubtless that the general notion of multiply extended magnitudes (in which space-magnitudes are included) remained entirely unworked. I have in the first place, therefore, set myself the task of constructing the notion of a multiply extended magnitude out of general notions of magnitude. It will follow from this that a multiply extended magnitude is capable of different measure-relations, and consequently that space is only a particular case of a triply extended magnitude. But hence flows as a necessary consequence that the propositions of geometry cannot be derived from general notions of magnitude, but that the properties which distinguish space from other conceivable triply extended magnitudes are only to be deduced from experience. Thus arises the problem, to discover the simplest matters of fact from which the measure-relations of space may be determined; a problem which from the nature of the case is not completely determinate, since there may be several systems of matters of fact which suffice to determine the measure-relations of space—the most important system for our present purpose being that which Euclid has laid down as a foundation. These matters of fact are—like all matters of fact—not necessary, but only of empirical certainty; they are hypotheses. We may therefore investigate their probability, which within the limits of observation is of course very great, and inquire about the justice of their extension beyond the limits of observation, on the side both of the infinitely great and of the infinitely small.

I. Notion of an *n*-ply extended magnitude

In proceeding to attempt the solution of the first of these problems, the development of the notion of a multiply extended magnitude, I think I may the more claim indulgent criticism in that I am not practised in such undertakings of a philosophical nature where the difficulty lies more in the notions themselves than in the construction; and that besides some very short hints on the matter given by Privy Councillor Gauss in his second memoir on Biquadratic

Residues, in the *Göttingen Gelehrte Anzeige*, and in his Jubilee-book, and some philosophical researches of Herbart, I could make use of no previous labors.

§1. Magnitude-notions are only possible where there is an antecedent general notion which admits of different specialisations. According as there exists among these specialisations a continuous path from one to another or not, they form a *continuous* or *discrete* manifoldness: the individual specialisations are called in the first case points, in the second case elements, of the manifoldness. Notions whose specialisations form a *discrete* manifoldness are so common that at least in the cultivated languages any things being given it is always possible to find a notion in which they are included. (Hence mathematicians might unhesitatingly found the theory of discrete magnitudes upon the postulate that certain given things are to be regarded as equivalent.) On the other hand, so few and far between are the occasions for forming notions whose specialisations make up a *continuous* manifoldness, that the only simple notions whose specialisations form a multiply extended manifoldness are the positions of perceived objects and colours. More frequent occasions for the creation and development of these notions occur first in the higher mathematic.

Definite portions of a manifoldness, distinguished by a mark or by a boundary, are called Quanta. Their comparison with regard to quantity is accomplished in the case of discrete magnitudes by counting, in the case of continuous magnitudes by measuring. Measure consists in the superposition of the magnitudes to be compared; it therefore requires a means of using one magnitude as the standard for another. In the absence of this, two magnitudes can only be compared when one is a part of the other; in which case also we can only determine the more or less and not the how much. The researches which can in this case be instituted about them form a general division of the science of magnitude in which magnitudes are regarded not as existing independently of position and not as expressible in terms of a unit, but as regions in a manifoldness. Such researches have become a necessity for many parts of mathematics, e.g., for the treatment of many-valued analytical functions; and the want of them is no doubt a chief cause why the celebrated theorem of Abel and the achievements of Lagrange, Pfaff, Jacobi for the general theory of differential equations, have so long remained unfruitful. Out of this general part of the science of extended magnitude in which nothing is assumed but what is contained in the notion of it, it will suffice for the present purpose to bring into prominence two points; the first of which relates to the construction of the notion of a multiply extended manifoldness, the second relates to the reduction of determinations of place in a given manifoldness to determinations of quantity, and will make clear the true character of an n-fold extent.

§2. If in the case of a notion whose specialisations form a continuous manifoldness, one passes from a certain specialisation in a definite way to another, the specialisations passed over form a simply extended manifoldness, whose true character is that in it a continuous progress from a point is possible only on two sides, forwards or backwards. If one now supposes that this manifoldness in its turn passes over into another entirely different, and again in a definite way, namely so that each point passes over into a definite point of the other, then all the specialisations so obtained form a doubly extended manifoldness. In a similar manner one obtains a triply extended manifoldness, if one imagines a doubly extended one passing over in a definite way to another entirely different; and it is easy to see how this construction may be continued. If one regards the variable object instead of the determinable notion of it, this construction may be described as a composition of a variability of $n + 1$ dimensions out of a variability of n dimensions and a variability of one dimension.

§3. I shall now show how conversely one may resolve a variability whose region is given into a variability of one dimension and a variability of fewer dimensions. To this end let us suppose a variable piece of a manifoldness of one dimension—reckoned from a fixed origin, that the values of it may be comparable with one another—which has for every point of the given manifoldness a definite value, varying continuously with the point; or, in other words, let us take a continuous function of position within the given manifoldness, which, moreover, is not constant throughout any part of that manifoldness. Every system of points where the function has a constant value, forms then a continuous manifoldness of fewer dimensions than the given one. These manifoldnesses pass over continuously into one another as the function changes; we may therefore assume that out of one of them the others proceed, and speaking generally this may occur in such a way that each point passes over into a definite point of the other; the cases of exception (the study of which is important) may here be left unconsidered. Hereby the determination of position in the given manifoldness is reduced to a determination of quantity and to a determination of position in a manifoldness of less dimensions. It is now easy to show that this manifoldness has $n - 1$ dimensions when the given manifoldness is n-ply extended. By repeating then this operation n times, the determination of position in an n-ply extended manifoldness is reduced to n determinations of quantity, and therefore the determination of position in a given manifoldness is reduced to a finite number of determinations of quantity *when this is possible*. There are manifoldnesses in which the determination of position requires not a finite number, but either an endless series or a continuous manifoldness of determinations of quantity. Such manifoldnesses are, for example, the possible determinations of a function for a given region, the possible shapes of a solid figure, &c.

II. Measure relations of which a manifoldness of *n* dimensions is capable on the assumption that lines have a length independent of position, and consequently that every line may be measured by every other

II. Measure relations of which a manifoldness of *n* dimensions is capable on the assumption that lines have a length independent of position, and consequently that every line may be measured by every other

Having constructed the notion of a manifoldness of n dimensions, and found that its true character consists in the property that the determination of position in it may be reduced to n determinations of magnitude, we come to the second of the problems proposed above, viz. the study of the measure-relations of which such a manifoldness is capable, and of the conditions which suffice to determine them. These measure-relations can only be studied in abstract notions of quantity, and their dependence on one another can only be represented by formulæ. On certain assumptions, however, they are decomposable into relations which, taken separately, are capable of geometric representation; and thus it becomes possible to express geometrically the calculated results. In this way, to come to solid ground, we cannot, it is true, avoid abstract considerations in our formulæ, but at least the results of calculation may subsequently be presented in a geometric form. The foundations of these two parts of the question are established in the celebrated memoir of Gauss, *Disquisitiones generales circa superficies curvas.*

§1. Measure-determinations require that quantity should be independent of position, which may happen in various ways. The hypothesis which first presents itself, and which I shall here develop, is that according to which the length of lines is independent of their position, and consequently every line is measurable by means of every other. Position-fixing being reduced to quantity-fixings, and the position of a point in the n−dimensioned manifoldness being consequently expressed by means of n variables $x_1, x_2, x_3, \ldots x_n$, the determination of a line comes to the giving of these quantities as functions of one variable. The problem consists then in establishing a mathematical expression for the length of a line, and to this end we must consider the quantities x as expressible in terms of certain units. I shall treat this problem only under certain restrictions, and I shall confine myself in the first place to lines in which the ratios of the increments dx of the respective variables vary continuously. We may then conceive these lines broken up into elements, within which the ratios of the quantities dx may be regarded as constant; and the problem is then reduced to establishing for each point a general expression for the linear element ds starting from that point, an expression which will thus contain the quantities x and the quantities dx. I shall suppose, secondly, that the length of the linear element, to the first order, is unaltered when all the points of this element undergo the same infinitesimal displacement, which implies at the same time that if all the quantities dx are increased in the same ratio, the linear element will vary also in the same ratio. On these suppositions, the linear element may be any homogeneous function of the first degree of the quantities dx, which is unchanged when we change the signs of all the dx, and in which the arbitrary constants are continuous functions of the quantities x. To find the simplest cases, I shall seek first an expression for manifoldnesses of $n - 1$ dimensions which are everywhere equidistant from the origin of the linear element; that is, I shall seek a continuous function of position whose values distinguish them from one another. In going outwards from the origin, this must either increase in all directions or decrease in all directions; I assume that it increases in all directions, and therefore has a minimum at that point. If, then, the first and second differential coefficients of this function are finite, its first differential must vanish, and the second differential cannot become negative; I assume that it is always positive. This differential expression, then, of the second order remains constant when ds remains constant, and increases in the duplicate ratio when the dx, and therefore also ds, increase in the same ratio; it must therefore be ds^2 multiplied by a constant, and consequently ds is the square root of an always positive integral homogeneous function of the second order of the quantities dx, in which the coefficients are continuous functions of the quantities x. For Space, when the position of points is expressed by rectilinear co-ordinates, $ds = \sqrt{\sum (dx)^2}$; Space is therefore included in this simplest case. The next case in simplicity includes those manifoldnesses in which the line-element may be expressed as the fourth root of a quartic differential expression. The investigation of this more general kind would require no really different principles, but would take considerable time and throw little new light on the theory of space, especially as the results cannot be geometrically expressed; I restrict myself, therefore, to those manifoldnesses in which the line-element is expressed as the square root of a quadric differential expression. Such an expression we can transform into another similar one if we substitute for the n independent variables functions of n new independent variables. In this way, however, we cannot transform any expression into any other; since the expression contains $\frac{1}{2}n(n+1)$ coefficients which are arbitrary functions of the independent variables; now by the introduction of new variables we can only satisfy n conditions, and therefore make no more than n of the coefficients equal to given quantities. The remaining $\frac{1}{2}n(n-1)$ are then entirely determined by the nature of the continuum to be represented, and consequently $\frac{1}{2}n(n-1)$ functions of positions are required for the determination of its measure-relations. Manifoldnesses in which, as in the Plane and in Space, the line-element may be reduced to the form $\sqrt{\sum dx^2}$, are therefore only a particular case of the manifoldnesses to be here investigated; they require a special name, and therefore these manifoldnesses in which the square of the line-element may be expressed as the sum of the squares of complete differentials I will call *flat*. In order now to review the true varieties of all the continua which may be represented in the assumed form, it is necessary to get rid of difficulties arising from the mode of representation, which is accomplished by choosing the variables in accordance with a certain principle.

§2. For this purpose let us imagine that from any given point the system of shortest lines going out from it is constructed; the position of an arbitrary point may then be determined by the initial direction of the geodesic in which it lies, and by its distance measured along that line from the origin. It can therefore be expressed in terms of the ratios dx_0 of the quantities dx in this geodesic, and of the length s of this line. Let us introduce now instead of the dx_0 linear functions dx of them, such that the initial value of the square of the line-element shall equal the sum of the squares of these expressions, so that the independent variables are now the length s and the ratios of the quantities dx. Lastly, take instead of the dx quantities $x_1, x_2, x_3, \ldots x_n$ proportional to them, but such that the sum of their squares $= s^2$. When we introduce these quantities, the square of the line-element is $\sum dx^2$ for infinitesimal values of the x, but the term of next order in it is equal to a homogeneous function of the second order of the $\frac{1}{2}n(n-1)$ quantities $(x_1\,dx_2 - x_2\,dx_1)$, $(x_1\,dx_3 - x_3\,dx_1)\ldots$ an infinitesimal, therefore, of the fourth order; so that we obtain a finite quantity on dividing this by the square of the infinitesimal triangle, whose vertices are $(0, 0, 0, \ldots)$, $(x_1, x_2, x_3, \ldots)$, $(dx_1, dx_2, dx_3, \ldots)$. This quantity retains the same value so long as the x and the dx are included in the same binary linear form, or so long as the two geodesics from 0 to x and from 0 to dx remain in the same surface-element; it depends therefore only on place and direction.

It is obviously zero when the manifold represented is flat, i.e., when the squared line-element is reducible to $\sum dx^2$, and may therefore be regarded as the measure of the deviation of the manifoldness from flatness at the given point in the given surface-direction. Multiplied by $-\frac{3}{4}$ it becomes equal to the quantity which Privy Councillor Gauss has called the total curvature of a surface. For the determination of the measure-relations of a manifoldness capable of representation in the assumed form we found that $\frac{1}{2}n(n-1)$ place-functions were necessary; if, therefore, the curvature at each point in $\frac{1}{2}n(n-1)$ surface-directions is given, the measure-relations of the continuum may be determined from them—provided there be no identical relations among these values, which in fact, to speak generally, is not the case. In this way the measure-relations of a manifoldness in which the line-element is the square root of a quadric differential may be expressed in a manner wholly independent of the choice of independent variables. A method entirely similar may for this purpose be applied also to the manifoldness in which the line-element has a less simple expression, e.g., the fourth root of a quartic differential. In this case the line-element, generally speaking, is no longer reducible to the form of the square root of a sum of squares, and therefore the deviation from flatness in the squared line-element is an infinitesimal of the second order, while in those manifold-nesses it was of the fourth order. This property of the last-named continua may thus be called flatness of the smallest parts. The most important property of these continua for our present purpose, for whose sake alone they are here investigated, is that the relations of the twofold ones may be geometrically represented by surfaces, and of the morefold ones may be reduced to those of the surfaces included in them; which now requires a short further discussion.

§3. In the idea of surfaces, together with the intrinsic measure-relations in which only the length of lines on the surfaces is considered, there is always mixed up the position of points lying out of the surface. We may, however, abstract from external relations if we consider such deformations as leave unaltered the length of lines—i.e., if we regard the surface as bent in any way without stretching, and treat all surfaces so related to each other as equivalent. Thus, for example, any cylindrical or conical surface counts as equivalent to a plane, since it may be made out of one by mere bending, in which the intrinsic measure-relations remain, and all theorems about a plane—therefore the whole of planimetry—retain their validity. On the other hand they count as essentially different from the sphere, which cannot be changed into a plane without stretching. According to our previous investigation the intrinsic measure-relations of a twofold extent in which the line-element may be expressed as the square root of a quadric differential, which is the case with surfaces, are characterized by the total curvature. Now this quantity in the case of surfaces is capable of a visible interpretation, viz., it is the product of the two curvatures of the surface, or multiplied by the area of a small geodesic triangle, it is equal to the spherical excess of the same. The first definition assumes the proposition that the product of the two radii of curvature is unaltered by mere bending; the second, that in the same place the area of a small triangle is proportional to its spherical excess. To give an intelligible meaning to the curvature of an n-fold extent at a given point and in a given surface-direction through it, we must start from the fact that a geodesic proceeding from a point is entirely determined when its initial direction is given. According to this we obtain a determinate surface if we prolong all the geodesics proceeding from the given point and lying initially in the given surface-direction; this surface has at the given point a definite curvature, which is also the curvature of the n-fold continuum at the given point in the given surface-direction.

§4. Before we make the application to space, some considerations about flat manifoldnesses in general are necessary; i.e., about those in which the square of the line-element is expressible as a sum of squares of complete differentials.

In a flat n-fold extent the total curvature is zero at all points in every direction; it is sufficient, however (according to the preceding investigation), for the determination of measure-relations, to know that at each point the curvature is zero in $\frac{1}{2}n(n-1)$ independent surface-directions. Manifoldnesses whose curvature is constantly zero may be treated as a special case of those whose curvature is constant. The common character of these continua whose curvature is constant may be also expressed thus, that figures may be moved in them without stretching. For clearly figures could not be arbitrarily shifted and turned round in them if the curvature at each point were not the same in all directions. On the other hand, however, the measure-relations of the manifoldness are entirely determined by the curvature; they are therefore exactly the same in all directions at one point as at another, and consequently the same constructions can be made from it: whence it follows that in aggregates with constant curvature figures may have any arbitrary position given them. The measure-relations of these manifoldnesses depend only on the value of the curvature, and in relation to the analytic expression it may be remarked that if this value is denoted by α, the expression for the line-element may be written

$$\frac{1}{1+\frac{1}{4}\alpha\sum x^2}\sqrt{\sum dx^2}\,.$$

§5. The theory of *surfaces* of constant curvature will serve for a geometric illustration. It is easy to see that surfaces whose curvature is positive may always be rolled on a sphere whose radius is unity divided by the square root of the curvature; but to review the entire manifoldness of these surfaces, let one of them have the form of a sphere and the rest the form of surfaces of revolution touching it at the equator. The surfaces with greater curvature than this sphere will then touch the sphere internally, and take a form like the outer portion (from the axis) of the surface of a ring; they may be rolled upon zones of spheres having less radii, but will go round more than once. The surfaces with less positive curvature are obtained from spheres of larger radii, by cutting out the lune bounded by two great half-circles and bringing the section-lines together. The surface with curvature zero will be a cylinder standing on the equator; the surfaces with negative curvature will touch the cylinder externally and be formed like the inner portion (towards the axis) of the surface of a ring. If we regard these surfaces as *locus in quo* for surface-regions moving in them, as Space is *locus in quo* for bodies, the surface-regions can be moved in all these surfaces without stretching. The surfaces with positive curvature can always be so formed that surface-regions may also be moved arbitrarily about upon them without *bending*, namely (they may be formed) into sphere-surfaces; but not those with negative curvature. Besides this independence of surface-regions from position there is in surfaces of zero curvature also an independence of *direction* from position, which in the former surfaces does not exist.

III. Application to Space

§1. By means of these inquiries into the determination of the measure-relations of an n-fold extent the conditions may be declared which are necessary and sufficient to determine the metric properties of space, if we assume the independence of line-length from position and expressibility of the line-element as the square root of a quadric differential, that is to say, flatness in the smallest parts.

First, they may be expressed thus: that the curvature at each point is zero in three surface-directions; and thence the metric properties of space are determined if the sum of the angles of a triangle is always equal to two right angles.

Secondly, if we assume with Euclid not merely an existence of lines independent of position, but of bodies also, it follows that the curvature is everywhere constant; and then the sum of the angles is determined in all triangles when it is known in one.

Thirdly, one might, instead of taking the length of lines to be independent of position and direction, assume also an independence of their length and direction from position. According to this conception changes or differences of position are complex magnitudes expressible in three independent units.

§2. In the course of our previous inquiries, we first distinguished between the relations of extension or partition and the relations of measure, and found that with the same extensive properties, different measure-relations were conceivable; we then investigated the system of simple size-fixings by which the measure-relations of space are completely determined, and of which all propositions about them are a necessary consequence; it remains to discuss the question how, in what degree, and to what extent these assumptions are borne out by experience. In this respect there is a real distinction between mere extensive relations, and measure-relations; in so far as in the former, where the possible cases form a discrete manifoldness, the declarations of experience are indeed not quite certain, but still not inaccurate; while in the latter, where the possible cases form a continuous manifoldness, every determination from experience remains always inaccurate: be the probability ever so great that it is nearly exact. This consideration becomes important in the extensions of these empirical determinations beyond the limits of observation to the infinitely great and infinitely small; since the latter may clearly become more inaccurate beyond the limits of observation, but not the former.

In the extension of space-construction to the infinitely great, we must distinguish between *unboundedness* and *infinite extent*, the former belongs to the extent relations, the latter to the measure relations. That space is an unbounded three-fold manifoldness, is an assumption which is developed by every conception of the outer world; according to which every instant the region of real perception is completed and the possible positions of a sought object are constructed, and which by these applications is for ever confirming itself. The unboundedness of space possesses in this way a greater empirical certainty than any external experience. But its infinite extent by no means follows from this; on the other hand if we assume independence of bodies from position, and therefore ascribe to space constant curvature, it must necessarily be finite provided this curvature has ever so small a positive value. If we prolong all the geodesics starting in a given surface-element, we should obtain an unbounded surface of constant curvature, i.e., a surface which in a *flat* manifoldness of three dimensions would take the form of a sphere, and consequently be finite.

§3. The questions about the infinitely great are for the interpretation of nature useless questions. But this is not the case with the questions about the infinitely small. It is upon the exactness with which we follow phenomena into the infinitely small that our knowledge of their causal relations essentially depends. The progress of recent centuries in the knowledge of mechanics depends almost entirely on the exactness of the construction which has become possible through the invention of the infinitesimal calculus, and through the simple principles discovered by Archimedes, Galileo, and Newton, and used by modern physic. But in the natural sciences which are still in want of simple principles for such constructions, we seek to discover the causal relations by following the phenomena into great minuteness, so far as the microscope permits. Questions about the measure-relations of space in the infinitely small are not therefore superfluous questions.

If we suppose that bodies exist independently of position, the curvature is everywhere constant, and it then results from astronomical measurements that it cannot be different from zero; or at any rate its reciprocal must be an area in comparison with which the range of our telescopes may be neglected. But if this independence of bodies from position does not exist, we cannot draw conclusions from metric relations of the great, to those of the infinitely small; in that case the curvature at each point may have an arbitrary value in three directions, provided that the total curvature of every measurable portion of space does not differ sensibly from zero. Still more complicated relations may exist if we no longer suppose the linear element expressible as the square root of a quadric differential. Now it seems that the empirical notions on which the metrical determinations of space are founded, the notion of a solid body and of a ray of light, cease to be valid for the infinitely small. We are therefore quite at liberty to suppose that the metric relations of space in the infinitely small do not conform to the hypotheses of geometry; and we ought in fact to suppose it, if we can thereby obtain a simpler explanation of phenomena.

The question of the validity of the hypotheses of geometry in the infinitely small is bound up with the question of the ground of the metric relations of space. In this last question, which we may still regard as belonging to the doctrine of space, is found the application of the remark made above; that in a discrete manifoldness, the ground of its metric relations is given in the notion of it, while in a continuous manifoldness, this ground must come from outside. Either therefore the reality which underlies space must form a discrete manifoldness, or we must seek the ground of its metric relations outside it, in binding forces which act upon it.

The answer to these questions can only be got by starting from the conception of phenomena which has hitherto been justified by experience, and which Newton assumed as a foundation, and by making in this conception the successive changes required by facts which it cannot explain. Researches starting from general notions, like the investigation we have just made, can only be useful in preventing this work from being hampered by too narrow views, and progress in knowledge of the interdependence of things from being checked by traditional prejudices.

This leads us into the domain of another science, of physic, into which the object of this work does not allow us to go to-day.

Synopsis

PLAN of the Inquiry:

I Notion of an n-ply extended magnitude

§1. Continuous and discrete manifoldnesses. Defined parts of a manifoldness are called Quanta. Division of the theory of continuous magnitude into the theories,

(1) Of mere region-relations, in which an independence of magnitudes from position is not assumed;
(2) Of size-relations, in which such an independence must be assumed,

§2. Construction of the notion of a one-fold, two-fold, n-fold extended magnitude.
§3. Reduction of place-fixing in a given manifoldness to quantity-fixings. True character of an n-fold extended magnitude.

II Measure-relations of which a manifoldness of n-dimensions is capable on the assumption that lines have a length independent of position, and consequently that every line may be measured by every other

§1. Expression for the line-element. Manifoldnesses to be called Flat in which the line-element is expressible as the square root of a sum of squares of complete differentials.
§2. Investigation of the manifoldness of n-dimensions in which the line-element may be represented as the square root of a quadric differential. Measure of its deviation from flatness (curvature) at a given point in a given surface-direction. For the determination of its measure-relations it is allowable and sufficient that the curvature be arbitrarily given at every point in $\frac{1}{2}n(n-1)$ surface directions.
§3. Geometric illustration.

§4. Flat manifoldnesses (in which the curvature is everywhere = 0) may be treated as a special case of manifoldnesses with constant curvature. These can also be defined as admitting an independence of n-fold extents in them from position (possibility of motion without stretching).

§5. Surfaces with constant curvature.

III Application to Space

§1. System of facts which suffice to determine the measure-relations of space assumed in geometry.

§2. How far is the validity of these empirical determinations probable beyond the limits of observation towards the infinitely great?

§3. How far towards the infinitely small? Connection. of this question with the interpretation of nature.

References

Clifford, William Kingdon. 1882. *Mathematical Papers by William Kingdon Clifford: Edited by Robert Tucker, with an introduction by HJ Stephen Smith*. Macmillan and Company.

Riemann, Bernhard. 1873. *On the Hypotheses Which Lie at the Bases of Geometry*, transl. William Kingdon Clifford. Nature VIII(183, 184): 14–17, 36, 37.

Riemann, Bernhard. 1959. On the hypotheses which lie at the bases of geometry. In *A Source Book in Mathematics*, ed. David E. Smith, 411–425 (transl. by Henry S. White, 1929). Mineola: McGraw-Hill.

Spivak, Michael. 1970–1975. *Differential Geometry*, vol. 1–5. Berkeley: Publish or Perish.

Presentation of the Text

4

4.1 Short Summary

In his work, Riemann analyzed the mathematical structure of space in a conceptually novel way. Through Riemann's work, physical space gets firstly empirically determinable characteristics and secondly loses its uniqueness as a mathematical space. For this purpose, Riemann first introduces the concept of a multiply extended variety or *manifold* (or *manifoldness* in Clifford's translation). A manifold is characterized in that its sufficiently small parts can be fully and non-redundantly described by n coordinates. That number n is then the dimension of the manifold. It is fundamental that this manifold structure determines in modern terminology only the topology, i.e. the qualitative aspects of position, but does not yet provide for any measurements. Riemann thus recognizes that in order to measure lengths and angles, an additional structure is required which is of a quantitative nature. This additional structure is arbitrary (obeying certain natural constraints). This structure can then be restricted on the one hand by conditions of simplicity and on the other hand by empirical testing if it is supposed to describe the actual physical space. Riemann then describes the quantitative structure by a so-called metric tensor,[1] which for simplicity is chosen as quadratic (this will be explained later). Using this metric tensor, one can then determine curve lengths and distances between points and sizes of angles, that is the usual metric quantities. But since a manifold can be described locally by coordinates in different ways, it becomes the central task of geometrical investigations to identify quantities that do not depend on the choice of coordinates. This then are the invariants of the manifold provided with a metric. Riemann thus goes on to identify a complete set of invariants under his conditions. This set of invariants is represented by the curvature tensor. This represents a far-reaching generalization of the Gaussian theory of surfaces. Through additional require-

[1] In Riemann's treatise, the concept of the tensor is not yet introduced, so that a subsequent development is anticipated by this formulation. That development is described in detail in Reich (1997).

© The Author(s), under exclusive license to Springer Nature Switzerland AG 2025
J. Jost, *Bernhard Riemann — On the Hypotheses Which Lie at the Bases of Geometry*, Classic Texts in the Sciences, https://doi.org/10.1007/978-3-031-91489-8_4

ments on the geometric properties, the curvature tensor can be more narrowly constrained. In particular, it follows from the requirement of the free mobility of rigid bodies that the curvature of the space has to be constant, a result which Helmholtz will later put at the center of his considerations. The Riemannian spaces of constant negative curvature turn out to be models of the non-Euclidean geometry of Bolyai and Lobatchevsky, as subsequently emphasized by Beltrami. Riemann therefore had found a new and much more general approach to non-Euclidean geometry, of which, incidentally, he had apparently not even been aware of when composing his work. For Riemann, this generality is particularly important from natural philosophical reasons because he already hints at the relationship, fundamental for Einstein's general theory of relativity, between the geometry of space and the forces caused by the objects contained in it. This extends far beyond the class of spaces of constant curvature, since then bodies moving in space affect the latter's geometry and then, conversely, the geometry can determine the motion of bodies.[2]

4.2 The Main Results of the Text

Riemann distinguishes between the qualitative manifold structure and the quantitative measurement structure, that is, between the topological and the metric structure of space, and developes the mathematical concepts needed for this purpose. The manifold structure refers only to the neighborhood structure and to the relative positions, i.e. to the qualitative aspects. The unboundedness of space, i.e., that it has no boundary, is an example of a topological property. For his concept of a manifold, Riemann assumes that the space can be locally described by coordinates, i.e., that it can be locally related to a (Cartesian) number space. This makes it possible to locally investigate a manifold with the tools of algebra and calculus. The number of independent coordinates that are necessary for this purpose then is the dimension n of the manifold. This dimension is not restricted to the number 3 of the realm of daily ordinary experience, but may take any value. As a result, the manifold concept also becomes a formal tool for the description of parameter-dependent structures of higher mathematics. Except for the requirements of independence and completeness that determine the dimension, the local coordinates describing space can be chosen arbitrarily. It is then the task of geometry to find invariants of a given manifold that are independent of such an arbitrary description.

A manifold can carry an additional structure, a quantitative metric structure that makes it possible to measure distances and angles. In order to obtain a concept that is sufficiently rich in content, Riemann assumes that this metric structure, when one looks at it infinitesimally,

[2] Pulte (2005), on the basis of his analysis of the natural philosophical and physical ideas of Riemann, however, is very critical with regard to the claim frequently expressed in the literature, for instance by Weyl in his commentary on Riemann's work [reproduced in Jost (2025a), Riemann (1990, 2013)], that Riemann had already intuitively guessed or anticipated important aspects of the general theory of relativity. See Sect. 4.5 below.

reduces to a Euclidean metric structure, so that infinitesimally the Pythagorean theorem applies. Locally, such a metric structure, however, generally deviates from the Euclidean one, which is exemplified by the fact that the sum of the angles in a triangle formed by geodesics does not have to be necessarily π. The deviation from the Euclidean structure is measured by the curvature of surfaces in space. From these curvatures, Riemann obtains a complete system of independent invariants to characterize the metric structure. Figures or bodies can be freely moved around in such a Riemannian manifold without stretching or compressing them precisely when the curvature is constant, that is, in every point and in every surface direction is the same. Among these spaces of constant curvature are the non-Euclidean geometries, which Riemann, however, does not discuss.

From the fact that the metric structure is an additional structure that is not included in the concept of a manifold, Riemann concludes that the metric of the space of our experience comes from outside, from physical forces. This anticipates the central idea of the general theory of relativity of Einstein, which identifies the curvature of space with the gravitational forces of the masses located or moving in it. Riemann and his successors, who formally elaborated and developed his geometric concepts, create the mathematical basis for the theory of general relativity with the principle of the independence of geometric relationships from their coordinate description and with the tensor calculus of Riemannian geometry.

For Riemann, mathematical space is the manifold, the multiply extended variety that can be represented in coordinates. The physical space of vision and touch, where we find the sensory objects, is an example, the color space another. This already exhausted the physical examples for Riemann. A mathematically profound and formative idea of Riemann is then that in mathematics, there are many such structures that can be considered as spaces. Here Riemann distinguishes two aspects, first, pure positional relationships and secondly the metric relations. The former leads into the realm of topology, still called Analysis Situs by Riemann (after Leibniz who had coined that term), for which he also created important foundations, whereas the latter leads into (Riemannian) geometry.

4.3 Riemann's Reasoning

The text consists of an introduction, in which the plan of study is presented, and three chapters which are sub-divided into paragraphs. The first chapter deals with the qualitative topological notion of a manifold, the second with the quantitative metric relations that can be given to a manifold, and then the third with applications to the (physical) space.

In the introduction, Riemann first discusses the relationship between nominal definitions which define the concept of space and specify the basic constructions in space on one hand, and axioms that contain the essential determinations on the other. It is not clear whether their relation is necessary nor whether it is possible.[3] In order to clarify this relationship,

[3] "Necessary" here probably means a necessity of thought in the Kantian sense, "possible" the logical possibility in the Leibnizian sense.

Riemann will first construct the concept of a multiply extended variety (manifold) in a general manner. This structure contains no metric, but only pure positional relationships, or in other words, the possibility of representing a point by specifying its coordinates by n real numbers. The metric relations can be obtained only empirically. These are facts that are not necessary, but only empirically certain, therefore, they are hypotheses—which explains his title.[4] Helmholtz will then write about the facts underlying geometry, as something fixed (the only empirically determined parameter that is left according to him is the value of the—for him, necessarily constant—curvature of space). Riemann, however, acknowledges the possibility of multiple systems that are sufficient to determine the metric properties of space, the Euclidean one being the most important. In particular, there is the question to what extent such a system can preserve its validity beyond the limits of observation, both small and large.

It may seem as somewhat surprising that Riemann considers empirical facts as hypotheses. The idea is that if the metric properties of the space do not necessarily follow from its structure, then the space can carry several possible metrics, and the mathematician then can specify any such hypothetical relations and examine the resulting structures and distinguish them with regard to their characteristics. Hilbert will then raise this as the axiomatic method to a systematic program.

After these preliminary considerations, the first part is devoted to the notion of the multiply extended variety, the manifold. The basis is "a general term … that allows various modes of determination," i.e., something that can be specified in different ways, that can assume different values. This concept constitutes the manifold, and its possible values provide the points or elements of this manifold. The discrete case where the manifold consists of elements that can be counted—in modern terminology, this would be a discrete set—requires no further explanation. The continuous case, where the values vary continuously and the parts can be measured, constitutes, however, the basic concept of the text. The values can have multiple independent degrees of freedom, and their number n is then the dimension

[4] However, in one of the posthumous philosophical fragments Riemann writes in the context of a discussion of the concept of causality and the positions of Kant and Newton "It is now customary to mean by a *hypothesis* everything that is added to the phenomena by thought" see Riemann (1892, S. 525) (or Riemann 1990, S. 557), my translation and emphasis. How much reflected the use of the word "hypothesis" by Riemann really is, is difficult to decide for me. It is the question of whether Riemann intended a reference to the relativization of the validity claim that Osiander in his unauthorized foreword to the work of Copernicus had produced by declaring astronomical theories as pure hypotheses without further claim to truth, to the claim of Kepler that he had created an astronomy without hypotheses, or to the statement of Newton "hypotheses non fingo" that did not resolve the difficulty about what is the cause for the physical ability of bodies to exert an attractive force on other bodies without spatial contact or a mediating medium (for a recent discussion within the history of ideas, see for example, Blumenberg 2007, S. 341–370). In any case, Riemann's cited quote corresponds to the view that in the described line of discourse finally had been formed, not without substantial and not completely eliminated resistances, namely that physics should uncover the mathematical laws underlying the observed phenomena without hypotheses about the nature of the bodies involved and that herein lies its claim to reality.

of the manifold. There are only a few real-life examples of this, according to Riemann only the locations of the sensory objects, that is, the possible positions of points in the sensory space—which have three degrees of freedom, the three dimensions of space—and the colors—where the determination of the number of degrees of freedom is no longer so obvious. One of the key insights of Riemann is the relevance of the concept of a manifold for higher mathematics. For example, Riemann himself, by the geometric interpretation of a multivalued function by means of a branched covering surface, the so-called Riemann surface, completely transformed and revolutionized the entire field of complex analysis and the theory of elliptic integrals. This made a conceptual synthesis of analytic, geometric and algebraic aspects possible that to this day has decisively shaped the further development of mathematics.[5] The concept of a manifold does not yet imply any determination of a measure, and thus no possibility to compare geometric quantities (objects in the manifolds, subsets of the manifold) independently of their position. Geometric quantities can therefore at first only be compared when one is a part of the other, and even then one can only say that the first is smaller than the other, but one cannot specify how much it is smaller. Without a device for measuring there is only the relation of containment; this results in set-theoretic topology, a branch of mathematics which acquired a foundational status in twentieth century mathematics. Riemann already recognizes the importance of such concepts for different areas of mathematics, and cites as an example the multivalent analytic functions. The concept of a manifold, however, is more subtle than it might seem from the foregoing. The position of a point in an n-dimensional manifold is described by specifying its coordinates. One will probably think here first about or in terms of the Cartesian coordinates in the three-dimensional Euclidean space, where the position of a point in space is described by three real numbers, which are located on three mutually orthogonal coordinate axes. But it is important to realize that herein several arbitrary conventions are hidden and additional structures are utilized. First of all, the Euclidean space contains no distinguished zero point or origin of the coordinates as the intersection of the three Cartesian coordinate axes. So this origin must be chosen arbitrarily, to determine the coordinates. With a different choice of the origin, one and the same point in space would get other coordinates. Similarly, the three coordinate directions are constrained only by the requirement of orthogonality, as they should be perpendicular to each other, and are otherwise arbitrary. Another choice of directions would again assign the same point in space different coordinate values. The choice of a unit on each coordinate axis, that is, the scale, is a mere convention. Finally, the requirement that the coordinate axes should be perpendicular is based on the possibility to measure angles. Here, thus, a metrical structure, the possibility of a measurement, is drawn into the picture, which, as Riemann points out, is not yet included in the concept of a manifold. If one does not assume any angle measurement, we can only specify that the three coordinate axes point in different directions. Likewise, the fact that the coordinate axes should be straight, assumes a notion, that of the straight line, that is not contained in the concept of a manifold as such.

[5] For a modern introduction see for instance (Jost 2006).

Take another instructive example: The earth's surface is a two-dimensional manifold that can be represented in idealized form by a spherical surface. On this sphere, the position of a point can be determined by specifying its longitude and latitude. Longitude and latitude thus are its coordinates. The curves of constant latitude are curves of constant distance from the poles, and the meridians are great circles passing through these poles. The null meridian by convention is fixed as the meridian through Greenwich in England. Not only that, but also the position of the poles on the spherical surface is a convention (on the globe, the poles are not determined geometrically, but kinematically, as the points of intersection with the axis of revolution). The distance from the poles as well as the notion of great circles (these are determined by the fact that the shortest paths on the surface of the sphere run along great circles) in turn require the possibility of measurements, and thus do not emerge out of the manifold concept.

Coordinates are thus a convenient means for a description of the position of points in a manifold, but require additional arbitrary rules and conventions. The points of the manifold are given independently of any coordinates. They can therefore be described by different sets of coordinates. This raises a problem. If the choice of coordinates is arbitrary, we can arbitrarily switch between different descriptions, and if the same object is presents itself quite differently, depending on the description, it seems that all invariant content is lost. However, the Riemannian geometry solves this problem. An object presents itself in a given description in a specific manner, but when changing the description this representation is transformed according to definite rules. What makes up the object thus are not its coordinate descriptions, but the transformation rules it experiences when the coordinate description changes. This is the basic principle of the theory of general relativity of Einstein, namely that the laws of physics are independent of specific coordinate descriptions, in the sense that they transform under a change of coordinates according to fixed rules. This is the principle of covariance—not invariance because the representation is precisely not invariant—and its universality explains the name of the theory. Physical phenomena are relative in the sense that they depend on the choice of a reference system, but satisfy general transformation rules under the transition to another reference system.

If all depended on the choice of the description, it might even be the case that the dimension n of the manifold depends on the choice of coordinates. This number n is the number of those coordinate values that are at least required for specifying a point in the given manifold. This means that we choose the coordinates independently, so that none of the coordinate values can be calculated as a function of other coordinate values of the same point, because those coordinate values that can be determined from others and are thus redundant may be omitted without impairing the complete determination of the point. It was then shown by Luitzen Egbertus Jan Brouwer (1881–1966) (Brouwer (1911)) that this requirement of the minimal number of independent coordinates already fixes the dimension n of a manifold. Therefore, the dimension of a manifold is independent of the choice of coordinates. Riemann determines this dimension inductively. From an n-fold extended manifold, we obtain an $(n+1)$-fold extended one by adding an additional degree of freedom, as one can pass from the two-dimensional Euclidean plane by the addition of a dimension to three-dimensional space. Conversely, when on an n-dimensional manifold one specifies a continuous function, one obtains $(n - 1)$-dimensional manifolds as its level sets, that is, as the subsets on which the function assumes a fixed value. Conversely, if you change this value continuously, it generates the original n-dimensional manifold as a single-parameter family of $(n - 1)$-dimensional manifolds. (Riemann points out that in this procedure, one encounters in general exceptional, singular, sets of smaller dimension than $n - 1$, as the level sets of a continuous function on an n-dimensional manifold need not all be $(n - 1)$-dimensional manifolds. For example, the curves of constant latitude on the two-dimensional sphere, i.e. the level sets of the distance to the North Pole, shrink to points at the poles, hence lose one dimension. A more detailed investigation of the relationship between such singularities and the global topology of the underlying manifold has become an important branch of mathematics of the twentieth century.)

Riemann also provides for the possibility of infinite dimensional manifolds, for example, the manifold of all functions on a given region. Such a function has an infinite number of degrees of freedom, namely its values at the infinitely many points of the region. This points forward to another important research branch of mathematics of the twentieth century, functional analysis.

Before we explain the idea of a Riemannian metric, we want to illustrate the problem once more in the example of curved surfaces in three-dimensional space as analyzed already by Gauss.

As explained, a manifold describes only the juxtaposition of points. The concept of a manifold constrains such a juxtaposition by the requirement that it can be locally mapped by coordinates onto a region in a Cartesian number space. Except for the dimension, this is not further determined, but arbitrary, and only continuity conditions must be guaranteed for a transition from one coordinate system to another. Globally, however, the manifold carries a topological structure, which in particular (apart from cases that are regarded as trivial in this context) prevents the whole manifold from being covered by a single coordinate system, also called a chart. The sphere is a clear and easily visualized example of a two-dimensional manifold. Parts of it can be represented in coordinate systems, as in the maps or charts of an atlas of the earth's surface, as we have already explained. The entire sphere can, however, not be so represented. In cartography therefore one uses a globe instead of the maps of an atlas. One can only compose the whole of the surface from the different charts, but one cannot capture it in a single chart. These are still purely topological aspects. The same applies to any other surface of the same topological type, that is, for all closed surfaces without holes, for example, ellipsoids or ovaloids. Likewise, closed surfaces of a different topological type, such as an annular surface, that is, the surface of a circular tube (called a *torus* in mathematical terminology), or a pretzel surface, cannot be captured by a single chart either. Here, the situation is even more complicated than for the sphere. An important insight, which also resulted from Riemann's considerations, not only those in geometry, but also those in complex analysis and elliptic integrals, which then led the theory of Riemann surfaces, is that the concept of a manifold already includes qualitative positional relationships, and that, consequently, different manifolds can be distinguished by different positional relationships. An important example may illustrate this: A closed curve in the Euclidean plane or on a sphere divides that surface into two parts; in the Euclidean plane, these two parts can also be distinguished from each other as the interior and the exterior of the curve. On an annular surface, however, there are closed curves, such as the generating curves that go around once and for which this is not the case, i.e., they do not decompose the surface into two parts. After Riemann this is expressed by saying that the connectivity of an annular surface is different from that of a sphere or a the plane. By such qualitative relationships surfaces of annular type can be topologically distinguished from spherical surfaces.

This is independent of a metric. Without the possibility of quantitative measurements, however, a spherical and an oval surface, for instance as purely topological objects, are not distinguishable from each other, because they can be mapped to each other in a reversible manner. In particular, they both share the same connectivity. That a sphere and an oval surface cannot be topologically distinguished from each other, is probably intuitively hard to grasp for the reason that we always visualize them as metric objects. By being visualized in three-dimensional Euclidean space and not as abstract objects, they always already carry a metric, the one induced by the ambient Euclidean space. Since we can measure the lengths of curves in Euclidean space, we can also measure lengths of curves lying on surfaces in Euclidean space. The distance between two points on a surface is then the shortest possible length of all the curves that connect these two points on the surface. The fact that we take into account only curves that run entirely on the surface makes the distance between the points on the surface greater than that measured in Euclidean space without such a constraint on the connecting curves. In Euclidean space, we can connect the two points by a straight line segment, and its length is then the Euclidean distance. Since the line segment does not typically lie on the surface, the distance on the surface is greater, because on the surface, the two points are then connected to each other only by curves that are all longer than the Euclidean line segment.

After this insertion which is hopefully useful for the geometric intuition we now turn to the second part of Riemann's text. Only through Riemann's conceptual analysis can we gain the full understanding of the above exposition of surfaces in space, precisely and perhaps somewhat paradoxically as he completely abstracts from the fact that a surface may be located in Euclidean space.[6] This of course builds on the distinction between external and internal geometry already taken by Gauss. Only the external geometry takes the position in

[6] Gauss had incidentally also pointed to the fact that the German language, in contrast to Latin, where there is only the term "superficies", and the Western European languages, where there is only the derived term "surface", distinguishes between "Fläche" (a two-dimensional manifold) and "Oberfläche"(a surface bounding a three-dimensionsal body). (Gauss to Schumacher, July 31, 1836 (Gauss 1973, Bd. 3, S. 164f) and September 3, 1842 (Gauss 1973, Bd. 4, S. 83f). I thank Rüdiger Thiele for this observation.) This is of course essential for the Gaussian surface theory, because in particular, he can speak about the bending of a surface without having to think at the same time about the deformation of a body.

space into account, while the internal geometry is solely concerned with metric relations on the surface itself.

This second part of Riemann's text now deals in a more abstract manner with the metric relations with which an n-dimensional manifold can be equipped. Mathematics will later develop the general concept of a metric space, i.e., that of a set in which the distance $d(P, Q)$ between any two points P and Q can be measured. This distance should be always positive, if P and Q are different, moreover symmetric in P and Q, that is, $d(P, Q) = d(Q, P)$, and, finally, the triangle inequality $d(P, Q) \leq d(P, R) + d(R, Q)$ should hold for any three points P, Q, R. The triangle inequality implies that the distance cannot decrease if an intermediate point is inserted. This is an axiomatic characterisation of a general distance concept. Riemann, however, proceeds differently and comes to the notion of what will be later named a Riemannian metric after him. He obtains his distance notion by measuring the lengths of curves. If one can measure the lengths of curves, the distance between two points is the length of the shortest curve joining them.[7] (In Euclidean space, this is the straight line connecting the two points concerned; in a general Riemannian space, this is called a geodesic curve.[8]) Riemann's notion of distance is thus a derived one, and assumptions about the determination of the length of curves lead Riemann to his metric concepts. The possibility of length determination naturally implies that each curve can be measured by each other, that is, that a length scale can be transported in the manifold, without changing its length. Curves are thereby considered as one-dimensional objects, and the length scale is therefore also a one-dimensional object, and not a rigid body. Helmholtz later demanded the free mobility of rigid bodies as a fundamental geometric fact. This requirement then leads necessarily to a much more specific form of geometry than Riemann's approach. More precisely, it implies that an n-dimensional space where an n-dimensional rigid body moves freely, is necessarily a Riemannian manifold of constant curvature. According to Riemann, this actually follows already from the assumption that two-dimensional figures are freely movable, without having to stretch, compress or distort them. The Riemann curvature concept will be explained below, but the essential point is that a general Riemannian manifold can have curvatures that vary from point to point and from surface direction to surface direction. Thus, the Riemannian approach is considerably more general than that of Helmholtz. This may initially be viewed as a disadvantage, insofar as Helmholtz unlike Riemann managed to determine the structure of the physical space completely from empirical facts (the still free curvature constant can also be in principle empirically determined by the sum of the angles in geodesic triangles), while the general Riemannian space has many contingent degrees of freedom. It turned then

[7] For mathematical correctness: It does not follow from the general concepts that it is always the case that for every two points on a manifold equipped with a Riemannian metric, there exists shortest connection between them. Assuming that there exists a connection at all (i.e., that the manifold is path-connected), it is possible to define the distance as the infimum of the lengths of all connecting curves.

[8] This name indicates the origin of modern differential geometry in the investigations of Gauß (1828) for land surveying.

out, however, that this is exactly the structure required for the general theory of relativity. In that theory, the curvature of space is determined by Einstein's equations through the gravitational forces of masses located in it, and conversely, precisely the degrees of freedom available in the Riemannian structure are needed to ensure that the gravitational forces can unfold.

Thus Riemann's approach is based on the possibility of invariant length measurements. This, however, appears to him to be too general (although mathematics has later investigated structures of such generality), and he is therefore looking for meaningful additional requirements. The first such requirement is that the length measurement is reduced to infinitesimal measurements, so that one measures the length of infinitesimal curve elements (we would today speak of tangent vectors) and then computes the length of a continuously differentiable curve by integrating these infinitesimal lengths along the curve. Riemann's conception thus finds its natural place in the context of mathematical analysis, differential and integral calculus.[9]

We want to express this still differently: A curve connects two points with each other, and it is ultimately the distance between these two points that is to be calculated. The analysis proceeds by considering at each point on the curve its direction, that is, its tangent vector, and determining the latter's length. Summation (integration) of these infinitesimal lengths over all points of the curve then yields its length. This greatly simplifies the task, because instead of two points we now need to consider just a point and the directional elements (tangents) at this point. The Riemannian metric is then the prescription according to which the length of a directional element at a point is determined. So there enter two different types of variables into the metric, the points of the manifold and the directional elements at these points. The dependence of the metric on the points of the manifold is arbitrary[10]—herein lies the universality of Riemann's concept. However, the metric is required by Riemann to be linearly homogeneous as a function of the directional elements. This means that, when the directional element in question is stretched or compressed by some factor, its length changes by the same factor. In addition, the length should not change when the direction is reversed, because the length of a curve must not depend on the direction in which it is traversed. Even under these restrictions there are still several possibilities, and Riemann then opts for the simplest, namely that the length is obtained as the square root of a quadratic expression in the possible displacement directions.[11] Riemann justifies this choice as follows: At a given point P on the manifold one would like to have a function that reconstructs the distance from P. This function should be differentiable. Since all other points have a positive distance

[9] S. Lie will later criticize this as an approach that is ill-suited for axiomatic purposes because it is not elementary. See the comments below on Lie's reworking of Helmholtz's approach.

[10] Except that its components must be differentiable functions (although Riemann did not state the precise differentiability assumptions explicitly, for the calculation of the Riemann curvature tensor, the second derivatives of the metric with respect to the point on the manifold are needed).

[11] The general case was taken up and developed in the Göttingen dissertation of Paul Finsler, *Über Kurven und Flächen in allgemeinen Räumen,* 1918 (reprinted in Finsler 2013). He thereby founded the research field of Finsler geometry. A modern textbook is Bao et al. (2012).

from P, therefore the function must assume its minimum value 0 in P. According to the rules of differential calculus, therefore its first derivatives must vanish in P. Furthermore, the second derivatives must be non-negative there, and Riemann then assumes that they are positive. In a first approximation, the requested function is therefore quadratic at P, i.e., it is essentially the square of the distance from P. The distance itself is therefore obtained as the square root of this quadratic function.

The requirement that the length element be obtained from the square root of a quadratic expression in the possible directions of displacement has the consequence that, infinitesimally, the Theorem of Pythagoras, and thus the rules of Euclidean geometry apply. (This raises the question whether therefore in the context of the Riemannian theory a special position of Euclidean geometry results. In particular, also the non-Euclidean geometries are described by using this method. In the further development of differential geometry, this then finds the expression that the tangent space at each point of a (differentiable) manifold carries a linear structure, so the methods of linear algebra apply. Thus, the tangent spaces of a Riemannian manifold then also carry a Euclidean metric structure. The tangent space at a point expresses the infinitesimal aspects of the geometry and it is therefore a tool for an approximate description of the local geometry. Euclidean geometry can therefore assume this task of an approximate description of the local Riemannian geometry particularly well, because it builds upon the linear structure of the Cartesian space; this was developed by Hermann Grassmann. That the Euclidean geometry is a useful tool of description, does not imply its conceptual priority over other geometries. Riemann himself does not even speak of a Euclidean structure, but refers to this possibility of approximation as flatness in the smallest parts.) The deviation from Euclidean geometry shows up only when moving from one point to another and finds its analytical expression in the dependence of the metric on the points of the manifold.

Riemann then examines how many degrees of freedom there exist for this dependency.

At each point there are as many independent displacement directions as the number n of dimensions the manifold has. Then there are $n(n+1)/2$ different products of these directions (because products are independent of the order of the factors). By transformations of the n coordinates, one can then produce n relations between these (i.e., n of these degrees of freedom come from the the choice of coordinates, and therefore do not contain coordinate-independent information about the metric structure). Hence $n(n+1)/2 - n = n(n-1)/2$ degrees of freedom are left, which then characterize the metric structure of the manifold. Riemann identifies these degrees of freedom with his curvature quantities and obtains a geometric description of a metric structure on a manifold. These curvature values are calculated from the second derivatives of the metric tensor with respect to the points of the manifold. They represent invariants of the Riemannian manifold, therefore coordinate independent quantities. From the first derivatives of the metric, in contrast, no invariants can be obtained.

If coordinates can be chosen arbitrarily, they can also be selected in the most convenient manner. That means that coordinates can be constructed in which geometric relationships are expressed in a particularly simple manner or where they show themselves most clearly.

Riemann employs this to his advantage and introduces special coordinates, which then were later called *(Riemann) normal coordinates* and which have become a very useful tool in the geometric tensor calculus. In these coordinates, starting from an arbitrarily selected reference point P the location of another point Q in its vicinity is described by its distance from P and the direction of the shortest connection from P to Q at P. In Euclidean space, this provides the well-known polar coordinates, and in a first approximation in a Riemannian space at the reference point P the metric looks like the Euclidean metric. In general, this is strictly true only for this point P itself, but because one can perform this construction at every point, this is sufficient for the intended purpose.

Why, then, have the coordinates chosen by Riemann so favorable properties? This rests first of all on the fact that in one dimension, there is no difference between Euclidean and Riemannian geometry. Each curve equipped with a measure is in itself, that is, *intrinsically*, indistinguishable from a Euclidean straight line. By virtue of a suitable choice of coordinates, the curve can be put into the Euclidean form. To do this, simply uniformly choose the coordinates adapted to the measure, i.e., such that equal distances on the curve correspond to the same coordinate differences. When one conceives the passage of the coordinate values as traversing through the curve, the curve is traversed in this way with constant speed, because the ratio of the length, measured along the curve, to the time, measured in the coordinates, remains constant. A piece of a curve therefore possesses no geometric invariants in itself, except its length. Curves thus do not differ in their intrinsic geometry from each other, but one and the same piece of a curve can only be described differently by different coordinates or parameterizations. It is the objective of geometry in the sense of Gauss and Riemann to exhibit geometric properties of objects that are independent of the chosen description.

Since, as explained, in one dimension, there are no intrinsic differences between a curve and a Euclidean straight line, this also applies to the shortest path from P to Q, a so-called geodesic curve in a Riemannian manifold. The only coordinate that this curve therefore contributes is its length, i.e., the distance between P and Q. Now this curve is not an arbitrary curve, but a geodesic one, a shortest connection. Like a straight line in the Euclidean plane, it therefore does not have any lateral deviation in the manifold, but steers from P directly to its target Q. Thus, it sits in the manifold like a straight line in the Euclidean plane. Here, too, if we move away from the inner geometry of the curve and consider its location in the ambient manifold, in the first approximation we cannot detect any differences to the Euclidean situation. In order to detect differences and thereby gain invariants, we have to move to a two-dimensional situation, a surface. According to the insights and findings of Gauss presented above, to which Riemann refers here, we know that a surface regardless of its position in an external surrounding space possesses an inner geometric invariant, its (Gauss) curvature. The idea of Riemann now consists in constructing a complete set of geometric invariants for a manifold at some point P from the intrinsic curvatures of different surfaces in that manifold. These surfaces can be obtained with the help of his coordinates discussed above. For this purpose, he considers the surfaces that consist of all geodesic curves emanating from P and whose directions at P lie in the same plane. Each infinitesimal plane at P , that is, the choice of two independent coordinate directions in P, therefore yields a surface in the manifold. The curvatures of these surfaces, called the sectional curvatures, at the point P then determine the geometry of the manifold at this point. Now, there are $n(n-1)/2$ independent plane directions in an n-dimensional space, and therefore Riemann obtains exactly the correct number of invariants in order to determine the geometry of a manifold in a unique and non-redundant manner.

This can also be understood geometrically as follows: We consider besides P not only one other point Q, but two others, Q and R, both of which have the same distance from P, and the shortest connections from P to Q and R. Then we can also examine the distance between Q and R. If we vary the common distance to P, that is, vary the points Q and R, but keep their directions from P fixed, then in Euclidean geometry, the distance between Q and R grows proportionally to their distance to P. On a curved surface this is no longer true. In the case of positive curvature, this distance is growing at a lower rate, while for negative curvature it grows faster (even exponentially). In the case of positive curvature, geodesic lines thus do not move apart as Euclidean straight lines at a linear rate, but like the great circles on the sphere finally even come together, whereas they diverge exponentially for negative curvature. The curvature also shows up in the comparison of area with Euclidean reference objects. In some Riemann manifold, take as a surface the circular disk of radius r formed by the geodesics up to the distance r emanating from P with initial directions lying in a fixed plane. The area of this surfaces then differs from the area of a Euclidean disc of the same radius, that is, from πr^2 by a fourth-order correction term which is proportional to the curvature of this plane direction at P.

For a better understanding of this issue and of the geometric meaning of curvature we shall now introduce a concept, which is not found in Riemann, but only in Christoffel, Ricci and Levi-Civita (1873–1941) in their further elaboration of the Riemannian theory. This is the concept of parallel transport.[12] In Euclidean space we can identify a direction at point A with the parallel direction in another point B, because after the corresponding Euclidean postulate or axiom, for each direction at A, there is exactly one parallel direction at B. Through the concept of parallelism we thus obtain a natural correspondence between the directions in two different points. We can therefore easily identify the infinitesimal geometry at A as given by the different directions in A with the infinitesimal geometry at B. And if we then identify the geometry at B with that at a third point C, and finally, that at C in turn with that at A, we recover the original geometry at A. This works in the sense that if we transport a specific direction at A to B, then to C and finally back to A, we again obtain our initial direction at A, and not another direction at the point A. Now, the Euclidean parallel postulate no longer holds in a Riemannian manifold, that is to a direction in a point P, we can no longer unequivocally assign a parallel direction at another point Q, such that the geodesics starting in corresponding directions are parallel to each other in a suitable sense. (Parallelism might mean here, as in the discussion of the non-Euclidean geometry, that the geodesic lines in question do not meet; but then, depending on the specific

[12] A historical treatment of the corresponding development can be found in Bottazzini (1999).

structure of the Riemannian manifold, none or an infinite number of such parallels might exist.) In a Riemannian manifold there is therefore no direct comparison possible between the geometric situations in different points P and Q. In fact, this is not surprising, because any relationship between P and Q should somehow depend on the points lying between them. This is the same as in physics where an instantaneous action at a distance between two points is unfortunately postulated in Newtonian physics, but nevertheless is conceptually unsatisfactory and therefore was then replaced in the theories of Faraday, Maxwell and Einstein by a field concept. However, initially there is a substantial difference between the physical transmission of an effect through a field and the geometric transport along the—or a—shortest connecting curve between two points. In a field, the effect spreads in all directions from P and can therefore also reach the point Q on all possible paths, whereas the process of parallel transport will take place along a specific path. In modern physics, the two concepts then later achieved a synthesis, which can be seen particularly well in the Feynman path integral approach.

In this sense, Hermann Weyl[13] took up the concept of parallel transport by Levi-Civita and generalized it, in order to eliminate, or better to derive from an infinitesimal concept, the comparison of magnitudes in different points that was unmotivated for him in Riemann's theory and to thereby develop a Riemannian geometry that is consistently based only on infinitesimal concepts and operations. By means of a so-called affine connection geometric relations in different points can be compared, and if such a connection respects the metric, then even magnitudes can be compared. Thus, there is no longer a remote comparison at a distance, but this is obtained by the integration of infinitesimal comparisons along curves.

In order therefore to explain the parallel transport in a Riemannian manifold, we again consider the Euclidean situation, but now under an infinitesimal point of view. To this end, we connect the points A and B by a straight line g. Along g we then have a distinguished direction, namely its own direction. We can then identify the initial direction of g at A with its final direction at B. This is evident, but the crucial insight is now that we can use this direction as the reference direction. We can in fact transport any direction (any vector) v in A along g into a direction at B, by requiring that during this transport, the length of v and the angle of v with the direction of g always remain constant and that during this transport process also v does not rotate around g. In principle, this transport process could even be performed along any curve between A and B, not only along the straight line g, but it is clear that the result of the transport of one and the same vector from A to B will then depend on the final direction of the curve at B. The straight line is distinguished from the other curves by the fact that its own direction along its course remains parallel to itself, because the straight line does not curve away from its own direction.

This infinitesimal transport principle can now be transferred to a Riemannian manifold. We connect the points P and Q in question by the—or more precisely, a (because there may be several)—shortest geodesic curve c. Again we use the own direction (tangential direction)

[13] Weyl (1918a, b) and also Weyl (1993). See also the historical analysis of Scholz (2001, 1999). The concept of a connection was developed further in particular by Ehresmann (1948, 1951).

of this curve as a reference direction, and then transport other tangent directions from P to Q, stipulating that their lengths and their angle with the tangential direction of c remain constant and that they should also not rotate about c. A geodesic curve in a Riemannian manifold, like a Euclidean straight line (whose generalization it represents because of this property), is distinguished by the fact that it does not bend away from its own direction, as otherwise it would run a detour and lose its shortest property. In this manner, the concept of parallel transport in a Riemannian manifold is found. However, the result of parallel transport now in general depends on the choice of a connecting curve, because as the example of the different great circles on the sphere, which connect the north pole to the south pole, shows there may be more than one such connection.

The main difference from the Euclidean case now is that for a transport from P to Q, then from Q to R, and finally again back from R to P, the end result when one has returned to P, will in general be different from the original direction that one had started with at P. This result also depends on the two points Q and R as well as on the connecting geodesics. Expressed more concisely, the result depends on the path traversed before returning to the starting point. It turns out that this path dependence of the parallel transport can be measured by the Riemann curvature.

Through these constructions, we can also offer an explanation why only second, but not first derivatives of the Riemannian metric can provide geometric invariants (the curvature is calculated from the second derivatives). The first derivative refers to the change from point to point, that is, expresses how the metric changes when one, for example, runs from P to Q. But now, as we have analyzed, the relationship between the geometric relations in two different points is not invariant, but must be established by additional constructions such as parallel transport. This is also reflected in the freedom of choice of coordinates. There is no correlation that needs to be respected and no invariant relationship between coordinates in different points, but the geometric relations in different points can be described independently in coordinates. On the other hand, of course, the geometric relations at a point can be compared with themselves, just as in the parallel transport along a closed triangle, we could compare the final result with the initial state. Infinitesimally, the return to a point along a closed path is expressed by second derivatives. In this way, the curvature calculated from second derivatives of the metric then provides geometric invariants, and as Riemann had concluded by counting the available degrees of freedom, as explained above, we then have found all the invariants of a Riemannian metric.

At this point, the following consideration naturally offers itself: In an axiomatic foundation of geometry, one could also directly start from the concept of parallel transport without the need for a metric. Parallel transport would then simply be a rule for identifying directions at two different points of a manifold along and depending on a connecting curve, with certain consistency requirements, which then lead to the axioms. Such a concept is also called a connection because it establishes the link or connection between the various points of a manifold. In particular, a connection permits a new metric-independent definition of geodesic curves, namely as those curves whose direction along itself always remains parallel. In this context, Kant's remark appears in a surprising new light, by which he supported his view that the proposition that the straight line (in Euclidean space) is the shortest connection between their endpoints, constitutes an example of a "synthetic a priori judgment": "For my concept of the straight contains nothing of quantity, but only a quality. The concept of the shortest is therefore entirely additional to it, and cannot be extracted out of the concept of the straight line by any analysis."[14]

This conceptual contrast, however, between straightness and shortest property had already been discussed long before Kant, since classical antiquity, where these two definition possibilities of a straight line passing, either through an inner quality or through an external metric were put forward by Euclid and Archimedes. Leibniz then analyzed in great detail the various determinations of a straight line and reached significant insights which, however, because not systematically published, could not influence the subsequent development.[15]

In mathematical terms, the facts are as follows: As the axiomatic notion of a connection shows, the concept of a straight line (in the sense of a geodesic curve) can be introduced by a purely infinitesimal concept, the self-parallelism of its tangent direction, without recourse to a distance and a shortest property. Conversely, in a Riemannian manifold, the condition that a curve be geodesic in the sense of being self-parallel can also be derived from the requirement that it represent the shortest connection between any two of its points. Only, there is no general reason that the two concepts of a geodesic curve, self-parallelism, i.e., straightness, and shortest property have to match. For the concept of the connection is designed such that it is not derived from a metric and therefore, in a specific case, it need not be derived from a metric. Metric and connection are logically independent concepts. Although a metric defines a particular connection (the so-called Levi-Civita connection), for which parallel transport leaves the metric relations unchanged, nevertheless, on a given manifold, also other connections can be introduced that satisfy all axioms required for this concept, but without respecting the metrical conditions. The geodesics of such a connection then no longer possess the shortest property.

After this digression, which we hope will be useful for a fuller understanding, we return now to the considerations of Riemann.

Riemann refers to those manifolds whose curvature is everywhere zero, as flat. He avoids at this point, however, to speak of a Euclidean structure, possibly because he had not taken note of the discussion about the the non-Euclidean geometries. Instead, he assigns the manifolds of vanishing curvature to the larger class of manifolds of constant curvature. (The non-Euclidean geometries of Gauss, Bolyai and Lobatchevsky are just the Riemannian geometries of constant negative curvature, while the geometries of constant positive curvature describe the spherical surfaces and the projective planes which are obtained from them by the identification of antipodal points, as well as their higher dimensional analogues. In

[14] Kant (1997, 2nd ed., B16). This argument is criticized, for instance, by Hegel (1986, p. 239f) in his *Science of Logic*.

[15] For a detailed exposition of Leibniz' reasonings, see De Risi (2007).

particular, Riemann, apparently without knowing the contemporary debate,[16] arrived by his own way at the non-Euclidean spaces. While these spaces for their creators provide alternatives to the Euclidean space that simply rest in themselves, for Riemann, in contrast, they emerge as special cases of a much more general theory that operates with general metric conditions and works in arbitrary dimensions.) Riemann then concludes that these spaces of constant curvature are precisely those in which figures can move without distortion. Since surfaces of different curvature differ in their inner geometric relations, the curvature must be the same at any point and in any two-dimensional direction at this point, for figures to move and rotate freely in space without thereby suffering any distortions. But since on the other hand, according to Riemann's considerations, the geometry is completely determined by the curvature, therefore also the geometry of a space of constant curvature has to be the same at every point and in each direction. Consequently, figures in such a space do not feel any difference caused by their position and thus can be moved freely. (Conversely, the free mobility of bodies was the starting point of geometric considerations of Hermann von Helmholtz which in the beginning were carried out without knowing about Riemann's theory, but which then led him to the constant curvature spaces.) Riemann also provides the formula for the metric of constant curvature a, which, incidentally, is the only real formula in his text. Finally, Riemann introduces geometric models to visualize the surfaces, i.e., the two-dimensional spaces, of constant curvature.

In the third and last part of his lecture, Riemann then turns his thoughts to physical space. A flat space is characterized by the fact that its curvature vanishes everywhere, which is equivalent to the fact that the sum of the angles in any triangle is exactly π (180°). Under the assumption that the shape of bodies is independent of their positions, which Riemann here attributes to Euclid, the curvature is constant and this then determines the sum of the angles in triangles.

He then distinguishes between discrete space structures in which in principle exact determinations are possible, and continuous ones, in which each measurement is necessarily fraught with uncertainty, so that for reasons of principle no completely exact determinations of the metric structure are possible.

He also points to the important conceptual distinction between unboundedness and infinite extension. The former simply means that the space has no boundary. In particular, the spherical surface is one example of an, although finite, but unbounded, two-dimensional space. (Today, one calls such a manifold closed.) Unboundedness is a purely topological property, independent of the metric structure. Infinity, i.e., infinite extension, on the other

[16] On this issue, see Scholz (1982). Riemann mentioned Legendre, probably referring to the statements derived from Legendre that without using the Euclidean parallel postulate one may deduce from the other axioms that the angle sum cannot exceed 180° in a triangle and that if there is a triangle in which this angle sum is exactly 180°, this also applies to all other triangles. (The latter is just the Euclidean case, in the non-Euclidean geometry, the sum of the angles in each triangle is always less than 180°.) These assertions of Legendre are among the precursors of non-Euclidean geometry, and as Scholz argues, the fact that Riemann mentions Legendre can only be understood if Riemann did not know the proper works on non-Euclidean geometry.

hand is a metric property because it means for example that one can move arbitrarily large distances away from any point.

The last paragraph contains Riemann's ideas about the physical causes for the metric properties of space. In a footnote at the end of his text, he states that this section still requires a revision and further elaboration. Thus, although Riemann's thoughts are here only very briefly sketched, he has nevertheless intuitively grasped significant aspects of the physics of the twentieth century. Based on the one hand on the mathematical methods of calculus and on the other hand on the experimental perspectives, which the microscope has opened, Riemann poses the question of the metric relations of space at the unmeasurably small scale, as he calls it. Although the independence of the bodies, i.e. the physical objects, from their position requires the constancy of the curvature of space, as Riemann has pointed out, the underlying empirical concepts of a solid body and a light beam seem to lose their validity in the infinitely small, so that the geometrical assumptions that he has made may no longer apply in that situation. One possibility is that space is ultimately discrete at a very small scale. Whether and to what extent this is the case is not yet finally resolved even in modern, contemporary physics. This results in the problems of quantum gravity, where the debates between various competing theories have not yet reached their decisive conclusion. In any case, with a purely discrete structure, we find ourselves in the realm of counting instead of measuring, so that the problem of an external justification for the metric structure here no longer poses itself for Riemann. In the case of a continuous spatial structure, however, according to Riemann, the reason for the basic metric relations must be sought outside, in binding forces acting thereon. Riemann therefore thinks of space as such only as a manifold without any further structure.[17] The additional structure of a Riemannian metric on the space is not a priori predefined, but is determined by physical forces. So, if these forces change, so do the metric properties of space. Physics does not take place in a given metric space, but as the spatial structure influences the course of the physical processes, so conversely, the physical forces by their effects shape space. In retrospect, this leads to the central idea of the general theory of relativity of Einstein, who in his field equations directly connects the curvature of space with the attractive forces of the masses contained in it, i.e., relates force to the curvature of space. It is of course a difficult and ultimately undecidable question of interpretation, how much of this Riemann really has already guessed. However, what cannot be denied is Riemann's ingenious intuition of the relationship between the metric structure of space and the physical forces acting in it or on it, i.e. the necessary deep connection between geometry and physics, on the basis of his novel conceptual analysis of spatial structure. In any case, Riemann and his successors provided the mathematical foundation for the general theory of relativity.

As stated, Riemann's text did not need or employ formulas. That, however, he was able to implement the presented conceptual considerations also algorithmically, he demonstrated in his 1861 Prize Essay submitted to the Paris Academy on heat propagation. But alas,

[17] This position has been called manifold realism, a special variant of structure realism, see e.g., Shapiro (2000).

this document did not meet a favorable fate. The prize was not granted to Riemann's essay because not all details of the proofs were provided. Accordingly, also this work was only posthumously published in the *Collected Works*,[18] after Riemann's successors Christoffel and Lipschitz had already developed a similar formalism (see below). Consequently, this work could not exert a profound influence. In the second edition of the collected works it was then extensively commented by the editor Heinrich Weber. Richard Dedekind had even worked out a still more extensive elaboration that anticipated some later developments, but likewise did not get[19] published either.

4.4 Mathematical Commentary

After the verbal and somewhat informal discussion of Riemann's argumentation and its context in the preceding section, I shall now turn to a more formal and rigorous mathematical treatment. In some parts, I shall not closely follow Riemann's reasoning, but rather present a systematic mathematical derivation that makes use also of insights developed by Riemann's successors Lipschitz, Christoffel, Ricci, Levi-Civita and Weyl, hoping to make Riemann's seminal ideas thereby clearer for a modern reader.[20] Of course, in contrast to the previous one, this section will presuppose some mathematical knowledge. In compiling this section, I have used Heinrich Weber's commentary on Riemann's Paris Academy Essay [pp. 405–423 of the second edition of Riemann's collected works (pp. 437–455 of the Narasimhan edition)], Hermann Weyl's commentary that is also reproduced in Riemann's collected works (see the bibliography in Chap. 7 for details), the detailed treatment of Michael Spivak (Spivak (1975, Vol. 2)) as well as my own Jost (2017) [of which I am preparing a new edition (Jost 2026)], and my German textbook written in collaboration with Jost Eschenburg (Eschenburg and Jost (2013)).

The treatment in the first two sections will be quite detailed and elementary, in order to make the subsequent less elementary considerations as accessible as possible.

---

[18] Commentatio mathematica, qua respondere tentatur quaestioni ab Illma Academia Parisiensi propositae: "Trouver quel doit être l'état calorifique d'un corps solide homogène indéfini pour qu' un système de courbes isothermes, à un instant donné, restent isothermes après un temps quelconque, de telle sorte que la température d'un point puisse s'exprimer en fonction du temps et de deux autres variables indépendantes", in Riemann (1892, pp. 423–436), with detailed comments by the editor, ibid. pp. 437–455. See also the analyses in Cogliati (2014), Darrigol (2015), Gray (2017).

[19] Now available in Sinaceur (1990); see also the discussion in Laugwitz (1996).

[20] A plausible putative reconstruction of Riemann's own computations, presumably building upon Gauß' *Disquisitiones* (Gauß 1828) is presened in Darrigol (2015).

4.4.1 The Concept of a Manifold

In a manifold of dimension n, the position of a point p is determined by the values of n variable scalar quantities $x^1, x^2, \dots, x^n$. The values $x^1(p), x^2(p), \dots, x^n(p)$ are called the coordinates of p. Different values of one or several of the x^i yield a different point. The choice of the coordinates is arbitrary, however. For different coordinates $y^1, y^2, \dots, y^n$, the values $y^1(p), y^2(p), \dots, y^n(p)$ for one and the same point p become different. Also, coordinates apply only locally, that is, in sufficiently small regions of the manifold, but not necessarily globally. Different coordinates may apply in different regions, but these regions may overlap. The important point is that in the vicinity of every point, suitable coordinates exist. [This is precisely the situation that arises when we represent parts of the globe (the surface of the earth), as an example of a manifold, by charts in an atlas. The same piece can be represented in different charts that overlap in that piece, but these charts can be of different scale, orientation, and so on. In particular, when we take the same Cartesian coordinates on each page of our atlas (in practice, in an atlas, one utilizes something coarser, a rectangular grid, and the positions of objects in a chart are provided in the index of the atlas only through the corresponding grid values), then one and the same point on the globe will have different coordinate values in different charts.]

This requires some topological precision, and the modern mathematical definition of a manifold is ascribed to Weyl (1913).[21] We do not enter into the details here, but need to point out at least the following. Riemann requires continuity, that is, the coordinate values $x^1(p), x^2(p), \dots, x^n(p)$ vary continuously when the point p varies continuously.[22] Later on, he also assumes differentiability, that is, the transition from one system of coordinates $x = (x^1, x^2, \dots, x^n)$ to another system $y = (y^1, \dots, y^n)$ is differentiable. Therefore, quantities depending on the point p of a manifold can unambiguously be differentiated in local coordinates. That is, when they are represented in one set of coordinates, like a function $F(p) = f(x^1(p), x^2(p), \dots, x^n(p))$, and if f is differentiable w.r.t. $x^1, \dots, x^n$, then the same applies to their representation in other coordinates. For instance, if we represent the same function F in other coordinates as $F(p) = g(y^1(p), \dots, y^n(p))$, then g is likewise a differentiable function of the coordinates $y^1, \dots, y^n$. In fact, if in the region of overlap of the two coordinate charts, we have the relation $x = \phi(y)$ (or with indices $x^i = \phi^i(y^1, \dots, y^n),\ i = 1, \dots, n$), then $g(y^1(p), \dots, y^n(p)) = f(\phi^1(p), \dots, \phi^n(p))$, and therefore by the chain rule

$$\frac{\partial g(y)}{\partial y^j} = \sum_{i=1}^{n} \frac{\partial f(x)}{\partial x^i} \frac{\partial \phi^i(y)}{\partial y^j}. \tag{4.1}$$

[21] For further references, see a footnote on p. 150.

[22] It then remains to specify what the latter means, a continuous variation of p, but in order to make this precise, one needs the concepts of set theoretical topology as developed by Hausdorff (1914). Riemann works with intuitive notions.

In particular, we see that for this to be valid, we need to require that the coordinate change $x = \phi(y)$ be differentiable.

Both $\frac{\partial g}{\partial y^j}$ and $\frac{\partial f}{\partial x^i}$ represent the derivatives of the function $F(p)$ in local coordinates. The values of these derivatives, however, are not the same, but are related by the transformation $\frac{\partial \phi^i}{\partial y^j}$. This is a fundamental principle. When we change coordinates (in a differentiable manner), then abstract properties like differentiability of a function are preserved, but the values of such objects change according to specific transformation rules.

Coordinate transformations need to be invertible. That is, when $x = \phi(y)$, then the transformation ϕ has an inverse $\psi = \phi^{-1}$, and $y = \psi(x)$. This is compatible with the transformation rule (4.1). The reason is that we have the relation

$$\left(\frac{\partial \psi^j}{\partial x^i}\right)_{i,j=1,\dots,n} = \left(\frac{\partial \phi^i}{\partial y^j}\right)^{-1}_{i,j=1,\dots,n} \tag{4.2}$$

for the matrices of the partial derivatives of a map and its inverse, or equivalently

$$\sum_{i=1}^{n} \frac{\partial \psi^j}{\partial x^i} \frac{\partial \phi^i}{\partial y^k} = \delta^j_k := \begin{cases} 1 & \text{for } j = k \\ 0 & \text{else.} \end{cases} \tag{4.3}$$

δ^j_k is called the Kronecker symbol [after Leopold Kronecker (1823–1891)]. Thus, with this relation, (4.1) is converted into

$$\frac{\partial f(x)}{\partial x^i} = \sum_{j=1}^{n} \frac{\partial g(y)}{\partial y^j} \frac{\partial \psi^j(x)}{\partial x^i}. \tag{4.4}$$

In a shorter, but more intuitive notation, we can write this as

$$\frac{\partial f(x)}{\partial x^i} = \sum_{j=1}^{n} \frac{\partial g(y)}{\partial y^j} \frac{\partial y^j}{\partial x^i}, \tag{4.5}$$

suggesting that simply a cancellation between the ∂y^j in the numerator and the denominator takes place. With this short-hand notation, the relation (4.3) becomes

$$\sum_{i=1}^{n} \frac{\partial y^j}{\partial x^i} \frac{\partial x^i}{\partial y^k} = \frac{\partial y^j}{\partial y^k} = \delta^j_k, \tag{4.6}$$

and this makes that relation obvious. In the sequel, we shall frequently use (4.6) without expliciting mentioning it.

The reference example against which we can view the concept of a manifold is $\mathbb{R}^n$ with its Cartesian coordinates $x^1, \dots, x^n$. There, we have a single, and apparently natural, coordinate chart. Nevertheless, it is sometimes expedient to also employ other coordinates, like the polar coordinates. On $\mathbb{R}^2$ with its Cartesian coordinates x^1, x^2, we write

$$x^1 = r\cos\varphi, \quad x^2 = r\sin\varphi; \tag{4.7}$$

we note that this coordinate change is *singular* at the origin $x^1 = x^2 = 0$, thus violating our general requirement of invertibility for coordinate changes, and therefore suitable care has to be exercised when extending computations to the origin. For later use, we state the formulae for the derivatives:

$$\frac{\partial x^1}{\partial r} = \cos\varphi, \quad \frac{\partial x^2}{\partial r} = \sin\varphi, \quad \frac{\partial x^1}{\partial\varphi} = -r\sin\varphi, \quad \frac{\partial x^2}{\partial\varphi} = r\cos\varphi. \tag{4.8}$$

Of course, polar coordinates can also be introduced in dimensions $n > 2$, by analogous, but somewhat more complicated formulae.

Let us consider another example. The *sphere* $S^n := \{(z^1, \dots, z^{n+1}) \in \mathbb{R}^{n+1} : \sum_{\alpha=1}^{n+1}(z^\alpha)^2 = 1\}$ is a manifold of dimension n. Charts can be given by stereographic projections from the north and the south pole onto the tangent plane of the opposite pole, that is, onto the planes $\{z^{n+1} = -1\}$, $\{z^{n+1} = 1\}$, resp. For that purpose, on $U_1 := S^n \backslash \{(0, \dots, 0, 1)\}$ we put

$$\begin{aligned} x(z^1, \dots, z^{n+1}) &:= (x^1(z^1, \dots, z^{n+1}), \dots, x^n(z^1, \dots, z^{n+1})) \\ &:= \left(\frac{2z^1}{1 - z^{n+1}}, \dots, \frac{2z^n}{1 - z^{n+1}}\right) \end{aligned} \tag{4.9}$$

as shown in the following figure (where $N = (0, \dots, 0, 1)$ is the north pole).[23]

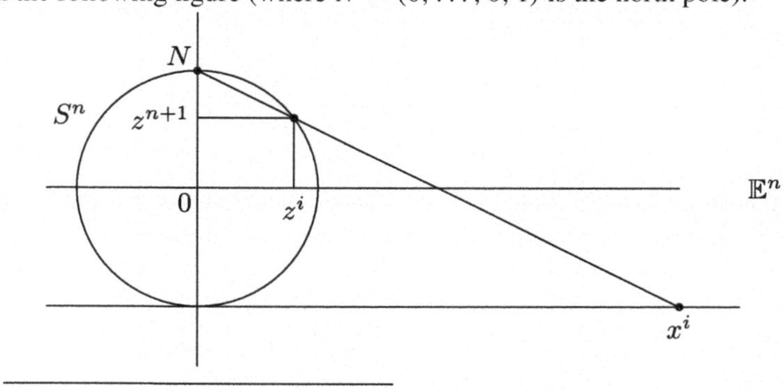

[23] We also see that we could have projected as well onto the equatorial plane $\mathbb{E}^n$, in which case we would have simply left out the factor 2 in the formulae for the x^i. We project here on the tangent plane of the south pole to conform with the formulae obtained below for constant curvature spaces.

Likewise, on $U_2 := S^n \backslash \{(0, \dots, 0, -1)\}$,

$$y(z^1, \dots, z^{n+1}) := (y^1(z^1, \dots, z^{n+1}), \dots, y^n(z^1, \dots, z^{n+1}))$$
$$:= \left(\frac{2z^1}{1 + z^{n+1}}, \dots, \frac{2z^n}{1 + z^{n+1}} \right). \tag{4.10}$$

Thus, we have two coordinate charts that overlap in the region $S^n \backslash \{(0, \dots, 0, 1), (0, \dots, 0, -1)\}$. The $z^1, \dots, z^{n+1}$ with the constraint $\sum_{\alpha=1}^{n+1} (z^\alpha)^2 = 1$ also parametrize the sphere, but they are not coordinates in the technical sense, because there are $n + 1$ of them, while the dimension of S^n is only n. Nevertheless, it is often obviously useful to carry out the computations in those z^α. For later purposes, we compute

$$1 = \sum_{\alpha=1}^{n+1} z^\alpha z^\alpha = \sum_{i=1}^{n} \frac{1}{4} x^i x^i (1 - z^{n+1})^2 + z^{n+1} z^{n+1}$$

hence

$$z^{n+1} = \frac{\frac{1}{4} \sum_{i=1}^n x^i x^i - 1}{\frac{1}{4} \sum_{i=1}^n x^i x^i + 1}$$

and then

$$z^i = \frac{x^i}{1 + \frac{1}{4} \sum_{i=1}^n x^i x^i} \quad (i = 1, \dots, n),$$

and then

$$\frac{\partial z^j}{\partial x^k} = \frac{\delta_{jk}}{1 + \frac{1}{4} \sum_{i=1}^n x^i x^i} - \frac{\frac{1}{2} x^j x^k}{(1 + \frac{1}{4} \sum_{i=1}^n x^i x^i)^2}$$
$$\text{for } j = 1, \dots, n, \ k = 1, \dots, n$$
$$\frac{\partial z^{n+1}}{\partial x^k} = \frac{x^k}{(1 + \frac{1}{4} \sum_{i=1}^n x^i x^i)^2}, \tag{4.11}$$

and by the chain rule, we could then also compute

$$\frac{\partial y^\ell}{\partial x^k} = \sum_\alpha \frac{\partial y^\ell}{\partial z^\alpha} \frac{\partial z^\alpha}{\partial x^k}.$$

In a simpler manner, from (4.9), (4.10), we have the coordinate transition formula

$$y^k = \frac{4x^k}{\sum_{i=1}^n x^i x^i},$$

and so

$$\frac{\partial y^k}{\partial x^j} = \frac{4\delta_{jk}}{\sum_{i=1}^n x^i x^i} - \frac{8x^j x^k}{(\sum_{i=1}^n x^i x^i)^2}. \tag{4.12}$$

Before proceeding, we should make the following general remark. While Riemann is clearly aware of the global aspects, his habilitation address is essentially concerned with purely local issues. Thus, for instance, the geometry on the surface of a cylinder is locally the same as that in the plane, because the latter can be rolled onto a cylinder without stretching or other deformations. Globally, or as one also says, topologically, these two surfaces are different. In the plane, every closed curve can be continuously contracted into a point, but this is not possible for a curve going around the cylinder. A space in which all closed curves can be contracted to points is called simply connected. While the cylinder itself thus is not simply connected, we can find sufficiently small regions on it that are. Thus, locally it is simply connected, like any manifold, and the considerations to follow can be confined to such regions. In this sense, we shall also always assume that the manifolds under consideration be connected, i.e., that any two points can be connected by a curve.

There is one comment on global aspects that Riemann does make. This concerns the distinction between *unboundedness* and *infinite extent*. Euclidean space is infinitely extended and has no boundary. The sphere, in contrast, is only of finite extension, but still does not have a boundary. Thus, unboundedness does not require infinite extent.

4.4.2 Tensor Calculus

The *tensor calculus* systematically incorporates relations like (4.6). It originated from the work of Riemann himself and Lipschitz and Christoffel and was fully developed by Ricci and turned out to be a most convenient tool for Einstein when working out his general theory of relativity. Although in its fully developed form, it is not yet present in Riemann's own work, I shall nevertheless employ it here for systematic reasons.[24]

The conventions of tensor calculus are carefully adapted to the transformation properties of the objects concerned, and because of this, it allows for rapid and automatic calculations. In that sense, it might be comparable to the ingenious notation that Leibniz had devised for the calculus, so that the chain rules for differentiation and integration are implemented algorithmically into the calculus.

As will become clear subsequently, in tensor calculus it is important to distinguish between upper and lower indices, as they indicate opposite transformation behavior. The coordinates carry upper indices, like x^i, and so do their infinitesimal versions, the differentials or covectors dx^i. In contrast, the index i of the vector $\frac{\partial}{\partial x^i}$ is considered as a lower index because it appears in the denominator below the fraction sign.

A basic convention then is that when the same index appears both in an upper and in a lower position in a product, it is to be contracted, and since this is a general rule,

[24] There also exists an alternative to Ricci's tensor calculus, the method of moving frames of Élie Cartan, but I shall not explain that here.

the summation indicating that contraction is usually left out. This is called the Einstein summation convention. Thus,

$$a^i b_i := \sum_{i=1}^{n} a^i b_i, \tag{4.13}$$

and this convention persists in the presence of other indices. For instance

$$c_{ik}a^i = \sum_{i=1}^{n} c_{ik}a^i \qquad \text{or} \qquad c_{ij}d^{ijk} = \sum_{i,j=1}^{n} c_{ij}d^{ijk}.$$

In the first example, only the index i appears twice in the product, whereas in the second example, both i and j, and hence both are to be summed over.

In principle, we could then also leave out the summation indices entirely and simply write ab or $c_k a, cd^k$. But the positions of the indices may remind us of the types of the tensors involved and therefore are not completely superfluous.

The Kronecker symbol introduced in (4.3) is particularly expedient; for instance, we have

$$\delta^i_k v^k = v^i. \tag{4.14}$$

Also, we can freely change the name of a summation index; thus, for instance

$$a^i b_i = a^k b_k. \tag{4.15}$$

A tangent vector then is a linear combination of the $\frac{\partial}{\partial x^i}$, whereas a covector is a linear combination of the differentials dx^i. That is, they are objects of the form

$$V = v^i \frac{\partial}{\partial x^i} \qquad \text{and} \qquad \omega = \omega_i dx^i, \tag{4.16}$$

with the summation convention applied (note the positions of the indices of the coefficients v^i and ω_i).

Let now $x = \phi(y)$, $y = \psi(x)$ as above, that is, consider a coordinate transformation. Then

$$\frac{\partial}{\partial x^i} = \frac{\partial}{\partial y^j}\frac{\partial y^j}{\partial x^i} \tag{4.17}$$

as in (4.5) or (4.6), and similarly

$$dx^i = \frac{\partial x^i}{\partial y^j} dy^j. \tag{4.18}$$

We note that in (4.17), the index i always appears in a lower position. Since we also want to have y^j in a lower position, we also have to have it in an upper position, in order to take a sum, and therefore we have to take the derivative of y w.r.t. x. In (4.18), this is the other way around, and we take the derivative of x w.r.t. y. In other words, vectors and covectors have the opposite transformation behavior. For the vector V in (4.16), this then means

$$V = v^i \frac{\partial y^j}{\partial x^i} \frac{\partial}{\partial y^j} =: w^j \frac{\partial}{\partial y^j}, \tag{4.19}$$

that is, the coefficients satisfy a so-called contravariant transformation rule,

$$w^j = v^i \frac{\partial y^j}{\partial x^i}. \tag{4.20}$$

For the covector ω in (4.16), we have the opposite, covariant transformation rule,

$$\omega = \omega_i \frac{\partial x^i}{\partial y^j} dy^j = \eta_j dy^j \tag{4.21}$$

with

$$\eta_j = \omega_i \frac{\partial x^i}{\partial y^j}. \tag{4.22}$$

Covectors and vectors are dual to each other in the sense that one can pair them to get a scalar quantity,

$$\omega(V) := v^i \omega_k \left(\frac{\partial}{\partial x^i}, dx^k \right) = v^i \omega_i \text{ with } \left(\frac{\partial}{\partial x^i}, dx^k \right) = \delta_i^k, \tag{4.23}$$

the Kronecker symbol from (4.3). The crucial point is that this is not affected by coordinate changes. In fact, according to (4.19)–(4.22), we have

$$\begin{aligned} w^j \eta_j &= v^i \frac{\partial y^j}{\partial x^i} \omega_i \frac{\partial x^i}{\partial y^\ell} \left(\frac{\partial}{\partial y^j}, dy^\ell \right) \\ &= v^i \omega_i \delta_\ell^j \left(\frac{\partial}{\partial y^j}, dy^\ell \right) \\ &= v^i \omega_i \delta_\ell^j \delta_j^\ell \\ &= v^i \omega_i \\ &= \omega(V). \end{aligned}$$

Thus, the opposite transformation behavior of vectors and covectors ensures that the result of their pairing, which is a scalar quantity, remains invariant under coordinate changes.

We now see the general principle. Objects like vectors or covectors are operators. Vectors can operate on functions by taking derivatives,

$$V(f) = v^i \frac{\partial f}{\partial x^i}, \tag{4.24}$$

or vectors and covectors can operate on each other by dual pairing.

There is an important difference between (4.23) and (4.24). Equation (4.23) depends only on the values of the objects involved, the vector V and the covector ω in this example, but not on their derivatives. In contrast, in (4.24), we take the derivative of the function f (but not of the vector V). Therefore, the operation (4.24) is not tensorial with respect to f, but only with respect to V. Later on, in Sect. 4.4.7, a crucial step will be the construction of tensors from such non-tensorial quantities by taking suitable differences.

The result of an operation as in (4.23) or (4.24) is a scalar quantity, that is, a number. The operational roles of V or ω do not depend on specific coordinates, but their representations (4.16) in those coordinates do. In order to leave the results of the operations invariant—numbers are numbers and should not depend on the representation—, the representations therefore need to transform appropriately under coordinate changes, according to the specific rules (4.20) and (4.22). In other words, one and the same geometric object looks different in different coordinate representations, and conversely, the transformation rules between different coordinate representations ensure the invariant character of the object. This is one of the fundamental insights that emerged from Riemann's concept of a manifold, even though the details of the calculus were only worked out by his successors. It is a fundamental principle in Einstein's theory of general relativity, called the *principle of covariance*. Note that it is not called the *principle of invariance*, as it expresses the fact that the representations do not stay invariant, but rather change according to general transformation rules.

We next analyze tensors with more than one index. In particular, when we discuss Riemannian metrics below, we shall need tensors of the form

$$A = a_{ij} dx^i dx^j. \tag{4.25}$$

Such an object operates on a pair of vectors $V = v^k \frac{\partial}{\partial x^k}$, $W = w^\ell \frac{\partial}{\partial x^\ell}$ to produce a scalar,

$$\begin{aligned} A(V, W) &= a_{ij} dx^i dx^j \left(v^k \frac{\partial}{\partial x^k}, w^\ell \frac{\partial}{\partial x^\ell} \right) \\ &= a_{ij} v^k w^\ell dx^i \left(\frac{\partial}{\partial x^k} \right) dx^j \left(\frac{\partial}{\partial x^\ell} \right) \\ &= a_{ij} v^k w^\ell \delta^i_k \delta^j_\ell \\ &= a_{ij} v^i w^j. \end{aligned} \tag{4.26}$$

Again, as we have seen in this computation, the positions of the indices incorporate the algorithmic rules. They also indicate the transformation behavior, which is

$$a_{ij} dx^i dx^j = a_{ij} \frac{\partial x^i}{\partial y^k} \frac{\partial x^j}{\partial y^\ell} dy^k dy^\ell. \tag{4.27}$$

Since the indices thus determine the type of a tensor, we shall often only write the coefficients to express a tensor; for instance, we shall simply speak of a_{ij} as a tensor, instead of writing $a_{ij} dx^i dx^j$.

We now turn to symmetries of tensors. A tensor with two indices might be symmetric or antisymmetric, that is,

$$a_{ji} = a_{ij} \qquad \text{or} \qquad b_{ji} = -b_{ij} \text{ for all } i, j. \tag{4.28}$$

From any such tensor d_{ij}, we can construct a symmetric or an antisymmetric one,

$$\frac{1}{2}(d_{ij} + d_{ji}) \qquad \text{or} \qquad \frac{1}{2}(d_{ij} - d_{ji}). \tag{4.29}$$

A tensor with three or more indices could have further symmetries, for instance

$$c_{ijk} + c_{jki} + c_{kij} = 0 \qquad \text{for all} \qquad i, j, k. \tag{4.30}$$

Products of tensors with opposite symmetries vanish. For instance, if $a_{ji} = a_{ij}$, $b^{ji} = -b^{ij}$, then

$$a_{ij}b^{ij} = a_{ji}b^{ij} = -a_{ji}b^{ji} = -a_{ij}b^{ij}, \tag{4.31}$$

where in the last step, we have simply renamed the indices in the summations, see (4.15). Thus, the expression in (4.31) has to vanish. We can also reformulate this observation in the following manner. If a_{ij} is a symmetric tensor, that is, $a_{ji} = a_{ij}$, then for any tensor c^{ij}, we have

$$c^{ij}a_{ij} = \frac{1}{2}\left(c^{ij} + c^{ji}\right)a_{ij}, \tag{4.32}$$

that is, we may assume that c^{ij} is likewise symmetric. By the same token, when b_{ij} is antisymmetric, that is, $b_{ji} = -b_{ij}$, we have

$$c^{ij}b_{ij} = \frac{1}{2}\left(c^{ij} - c^{ji}\right)b_{ij}, \tag{4.33}$$

that is, we may now assume that c^{ij} is antisymmetric.

Analogously, when the tensor c_{ijk} satisfies (4.30), and if we form a product

$$d^{ijk}c_{ijk}, \tag{4.34}$$

then we may assume that d^{ijk} also satisfies (4.30), that is, becomes 0 under cyclic permutations of the indices. In fact, we have, by renaming indices again,

$$(d_{ijk} + d_{jki} + d_{kij})c^{ijk} = d_{ijk}(c^{ijk} + c^{jki} + c^{kij}) = 0.$$

Thus, the simple trick of renaming indices allows us to transfer a symmetry from one factor to the other in a product. We shall use this frequently in Sect. 4.4.5 below.

4.4.3 Metric Structures

The next step consists in introducing a metric structure. The concept of a manifold does not yet contain any quantitative notions, and in fact, the preceding might even suggest that no quantitative notions are possible because all objects transform under coordinate changes. Fortunately, however, quantities are numbers, and as we have emphasized in the preceding, numbers should stay invariant after all. In particular, distances or lengths of curves are such numerical quantities.

Now, according to Riemann, measurements should be independent of the position where they are taken. This is a general principle, and Riemann then makes some additional assumptions to make it more specific. First, the lengths of curves in different positions should be comparable, that is, we should be able to measure the length of one curve through another one. In other words, he requires that line elements, that is, one-dimensional objects can be carried around freely in the manifold as measuring devices. He does not assume the same for higher-dimensional objects. Rather, the free mobility of rigid bodies in space is an additional requirement that is satisfied only for Riemannian manifolds of constant curvature, as we shall see below.

He then considers differentiable curves, that is, he lets the coordinates $x^1(t), \ldots, x^n(t)$ depend differentiably on a scalar parameter t. He then seeks a line element ds for infinitesinal length measurement, that is, in modern terminology, he wishes to measure the length of the tangent vector $\dot{x}(t) = \frac{dx}{dt}$ at some point p on the curve. The length element should then be a function of the position p and of the directions at that point. More precisely, since we want to obtain a scalar quantity from directions, the length element should contain the covectors $dx^1, \ldots, dx^n$ at p, according to the principles set forth in Sect. 4.4.2. Also, when we scale a line element, its length should scale by the same factor, and therefore, the length element should be a homogeneous function of the first degree of those dx^i.

Next, he invokes the $(n-1)$-dimensional manifolds that have constant distance from p in the n-dimensional manifold under consideration. Since distances are measured in terms of lengths, he seeks a function that expresses that distance. When one requires that this function be 0 at p and be positive elsewhere and furthermore differentiable, then its first derivatives at p vanishes and its second derivatives are nonnegative, and he assumes that they are positive. The natural such function is the square $f(y) = \text{dist}^2(p, q)$, where x stands for the coordinates of the variable point q, of the distance from p whose coordinates we suppose to be x_0. Its second derivatives then yield the tensor

$$g_{ij}(x_0)dx^i dx^j := \frac{1}{2}\frac{\partial^2}{\partial x^i \partial x^j} f(x)dx^i dx^j \text{ at } x = x_0, \text{ that is, at the point } p. \tag{4.35}$$

Such an expression $g_{ij}(x)dx^i dx^j$ is called a *Riemannian metric*. It is the fundamental object of Riemannian geometry.

We record the basic requirements:

1. The metric tensor is symmetric, that is, $g_{ji} = g_{ij}$.
2. It is positive definite, that is, $g_{ij} v^i v^j > 0$ whenever $(v^1, \ldots, v^n) \neq 0$.
3. $g_{ij}(x)$ depends twice differentiably on the coordinate x (this property will be needed to define the curvature).

The line element then is

$$ds = \sqrt{g_{ij}(x_0) dx^i dx^j}, \tag{4.36}$$

and for the special case of Euclidean space, this reduces to

$$ds = \sqrt{\sum (dx^i)^2}. \tag{4.37}$$

He briefly contemplates other possibilities, like taking the fourth root of a quartic expression, instead of the square root of a quadratic one as in (4.36), but he then proceeds with (4.36) as being the simplest.

Since second derivatives commute for a smooth function, the tensor g_{ij} in (4.35) is symmetric, $g_{ji} = g_{ij}$, as already stipulated above. It therefore has $\frac{n(n+1)}{2}$ independent components. Since, however, we have n coordinate degrees of freedom to transform that tensor, only $\frac{n(n+1)}{2} - n = \frac{n(n-1)}{2}$ degrees of freedom have independent geometric content. These geometric degrees of freedom will subsequently be identified as sectional curvatures.

In particular, in general, there is no coordinate transformation that reduces a given metric tensor $g_{ij}(x)$ at all points x simultaneously to the Euclidean form (4.37). Riemann calls a metric tensor whose line element is of the form (4.37) *flat*, and the flat metrics therefore are particular Riemannian metrics. They are characterized by the fact that all their sectional curvatures vanish.

Let

$$g_{ij} dx^i dx^j \quad \text{with } g_{ji} = g_{ij} \tag{4.38}$$

thus be a metric tensor. We can now introduce further conventions for tensor calculus. First of all, we denote the inverse of g_{ij} as a matrix by g^{ij}. That g_{ij} and g^{ij} are inverses of each other is expressed by the relation

$$g^{ij} g_{jk} = \delta^i_k. \tag{4.39}$$

(Note that the notation would become more systematic if we wrote g^i_k in place of δ^i_k.)

We then introduce the following conventions for raising or lowering indices of tensors.

$$v^i = g^{ij} v_j \quad \text{and} \quad v_i = g_{ij} v^j. \tag{4.40}$$

Also, as in (4.26), we can utilize a more abstract notion for the metric product of two vectors and put

$$g(V, W) = g_{ij} v^i w^j. \tag{4.41}$$

The behavior of the metric tensor under coordinate transformations will be the main object of the subsequent considerations. As an example, let us represent the Euclidean metric $dx^1dx^1 + dx^2dx^2$ of $\mathbb{R}^2$ in polar coordinates r, φ (4.7). We obtain, using (4.8),

$$\begin{aligned}(dx^1)^2 + (dx^2)^2 &= \left(\left(\frac{\partial x^1}{\partial r}\right)^2 + \left(\frac{\partial x^2}{\partial r}\right)^2\right) dr^2 \\ &\quad + 2\left(\frac{\partial x^1}{\partial r}\frac{\partial x^1}{\partial \varphi} + \frac{\partial x^2}{\partial r}\frac{\partial x^2}{\partial \varphi}\right) dr d\varphi \\ &\quad + \left(\left(\frac{\partial x^1}{\partial \varphi}\right)^2 + \left(\frac{\partial x^2}{\partial \varphi}\right)^2\right) d\varphi^2 \\ &= dr^2 + r^2 d\varphi^2. \qquad (4.42)\end{aligned}$$

From (4.11), we can also compute the metric on the sphere S^n in our coordinates $x^1, \ldots, x^n$ as

$$\begin{aligned}\sum_{\alpha=1}^{n+1} dz^\alpha dz^\alpha &= \left(\delta_{k\ell}\frac{\partial z^k}{\partial x^i}\frac{\partial z^\ell}{\partial x^j} + \frac{\partial z^{n+1}}{\partial x^i}\frac{\partial z^{n+1}}{\partial x^j}\right) dx^i dx^j \\ &= \frac{\delta_{ij}}{(1 + \frac{1}{4}\sum (x^i)^2)^2} dx^i dx^j \qquad (4.43)\end{aligned}$$

where the indices k, ℓ run from 1 to n. In the y-coordinates of (4.10), the metric looks the same.

4.4.4 Geodesic Curves

Whereas in the preceding section, we have followed Riemann's text closely, we now turn again to a systematic treatment, in order to develop the framework within which Riemann's ideas can be analyzed in formal terms. In any case, the constructions to follow were known since the work of Euler, Lagrange, and Jacobi, and so were also familiar to Riemann.

Let $[a, b]$ be a closed interval in $\mathbb{R}$, $c : [a, b] \to M$ a differentiable curve. We abbreviate the coordinate representation $(x^1(c(t)), \ldots, x^n(c(t)))$ as $(x^1(t), \ldots, x^n(t))$, as in the previous section, and put

$$\dot{x}^i(t) := \frac{d}{dt}(x^i(t)).$$

The *length* of c then is defined as

$$L(c) = \int_a^b \sqrt{g_{ij}(x(t))\dot{x}^i(t)\dot{x}^j(t)}\, dt. \qquad (4.44)$$

By the chain rule, the length of c is unaffected by reparametrizations, that is, if $\kappa : [\alpha, \beta] \to [a, b]$ is differentiable with nonvanishing derivative, then

$$L(c \circ \kappa) = L(c). \tag{4.45}$$

We want to investigate shortest curves, for instance, shortest connections between two points p and q in M. That means that the curve $c : [a, b] \to M$ with $c(a) = p, c(b) = q$ satisfies

$$L(c) \leq L(\gamma) \text{ for all } \gamma[a, b] \to M \text{ with } \gamma(a) = p, \gamma(b) = q. \tag{4.46}$$

In particular, we can compare the length of c with that of other curves in its vicinity, that is, with curves γ_ϵ of the form

$$x(t) + \epsilon\xi(t) \text{ with } \xi(a) = 0 = \xi(b) \tag{4.47}$$

in our local coordinates, for some ϵ of small absolute value $|\epsilon|$. In particular, the curves γ_ϵ constitute a variation of c, since $\gamma_0 = c$, and $\gamma_\epsilon(a) = p, \gamma_\epsilon(b) = q$ for all ϵ. Thus, since c as the shortest such curve minimizes the length L, we have

$$L(\gamma_0) \leq L(\gamma_\epsilon) \text{ for all } \epsilon. \tag{4.48}$$

Therefore, at such a minimum,

$$\frac{d}{d\epsilon} L(\gamma_\epsilon)_{|\epsilon=0} = 0. \tag{4.49}$$

Definition 4.1 A curve c that satisfies (4.49) for all such variations (4.47) is called a *geodesic*.

In particular, shortest, that is, length minimizing, curves are geodesics, because we have derived (4.49) from the length minimizing property.

We now want to derive the differential equation that a geodesic has to satisfy from the relation (4.49). The structure of the reasoning becomes clearer when we consider a more general situation, that is, an integral of the form

$$I(c) = \int_a^b f(\dot{x}(t), x(t))dt, \qquad \text{again with the abbreviation } x(t) = x(c(t)). \tag{4.50}$$

For our example, from (4.44) of course

$$f(\dot{x}(t), x(t)) = \sqrt{g_{ij}(x(t))\dot{x}^i(t)\dot{x}^j(t)}. \tag{4.51}$$

As in (4.49), we assume

$$0 = \frac{d}{d\epsilon} I(\gamma_\epsilon)_{|\epsilon=0} = \frac{d}{d\epsilon} \int_a^b f(\dot{x}(t) + \epsilon\dot{\xi}(t), x(t) + \epsilon\xi(t))dt_{|\epsilon=0} \tag{4.52}$$

for all variations as in (4.47). This implies

$$0 = \int_a^b \Big(f_{\dot{x}^i}(\dot{x}(t), x(t))\dot{\xi}^i(t) + f_{x^i}(\dot{x}(t), x(t))\xi^i(t)\Big)dt, \tag{4.53}$$

where a subscript like $\dot{x}^i$ indicates the argument of f with respect to which a partial derivative is taken. Integrating (4.53) by parts and using that $\xi(a) = 0 = \xi(b)$, we get

$$0 = \int_a^b \Big(- \frac{d}{dt} f_{\dot{x}^i}(\dot{x}(t), x(t)) + f_{x^i}(\dot{x}(t), x(t))\Big)\xi^i(t)dt. \tag{4.54}$$

Since this holds for all variations $\xi(t)$, we conclude that

$$- \frac{d}{dt} f_{\dot{x}^i}(\dot{x}(t), x(t)) + f_{x^i}(\dot{x}(t), x(t)) = 0 \text{ for all } i, \tag{4.55}$$

or writing the derivative with respect to t out,

$$f_{\dot{x}^i\dot{x}^k}(\dot{x}(t), x(t))\ddot{x}^k(t) + f_{\dot{x}^i x^k}(\dot{x}(t), x(t))\dot{x}^k(t) = f_{x^i}(\dot{x}(t), x(t)) \text{ for all } i. \tag{4.56}$$

The equations (4.56) are called the Euler-Lagrange equations for the variational problem given by the variational integral I from (4.50). They are named after Leonhard Euler who first derived them by a different method and Joseph Louis Lagrange (1736–1813) who found the elegant derivation that we have presented here.

Returning to our geodesic problem and inserting (4.51) into (4.55) yields

$$\begin{aligned}
&\frac{1}{f}\Big(g_{ij}(x(t))\ddot{x}^j(t) + \frac{\partial}{\partial x^k} g_{ij}(x(t))\dot{x}^k(t)\dot{x}^j(t)\Big) - \frac{1}{f^2}\frac{df}{dt} g_{ij}(x(t))\dot{x}^j(t) \\
&= \frac{1}{2f}\frac{\partial}{\partial x^i} g_{kj}(x(t))\dot{x}^k(t)\dot{x}^j(t) \\
&\text{for} i = 1, \ldots, n.
\end{aligned} \tag{4.57}$$

In order to simplify this system of equations, we assume that

$$f(\dot{x}(t), x(t)) = \sqrt{g_{ij}(x(t))\dot{x}^i(t)\dot{x}^j(t)} \equiv \text{const}. \tag{4.58}$$

This simply means that the length of the tangent vector $\dot{x}(t)$ is constant, that is, the curve $x(t)$ is traversed with constant speed. This can always be achieved by a reparametrization as in (4.45), that is, we can choose $\kappa(\tau)$ such that $\sqrt{g_{ij}(x(\kappa(\tau)))\frac{dx^i(\kappa(\tau))}{d\tau}\frac{dx^j(\kappa(\tau))}{d\tau}} \equiv \text{const}$, as long as $\frac{dx(t)}{dt} \neq 0$. Of course, when $\frac{dx(t)}{dt} \equiv 0$, the curve is constant and its length is 0. When this derivative vanishes only at some points or in some intervals, the reasoning needs to be a little more careful, but we suppress this technical issue here.[25] Thus, we assume that (4.58) is valid. Then $\frac{df}{dt} \equiv 0$ and (4.57) simplifies:

$$g_{ij}(x(t))\ddot{x}^j(t) + \frac{\partial}{\partial x^k} g_{ij}(x(t))\dot{x}^k(t)\dot{x}^j(t) - \frac{1}{2}\frac{\partial}{\partial x^i} g_{kj}(x(t))\dot{x}^k(t)\dot{x}^j(t) = 0. \tag{4.59}$$

We write

$$\frac{\partial}{\partial x^k} g_{ij}(x(t))\dot{x}^k(t)\dot{x}^j(t) = \frac{1}{2}\left(\frac{\partial}{\partial x^k} g_{ij}(x(t))\dot{x}^k(t)\dot{x}^j(t) + \frac{\partial}{\partial x^j} g_{ik}(x(t))\dot{x}^j(t)\dot{x}^k(t)\right)$$

utilizing the rules explained in Sect. 4.4.2, see (4.32). We insert this into (4.59) and use the abbreviation

$$g_{j\ell,k} = \frac{\partial}{\partial x^k} g_{j\ell}$$

to obtain

$$g_{ij}(x(t))\ddot{x}^j(t) + \frac{1}{2}(g_{ij,k} + g_{ik,j} - g_{kj,i})\dot{x}^k(t)\dot{x}^j(t) = 0.$$

We then multiply by $g^{\ell i}$ to get

$$\ddot{x}^\ell(t) + \Gamma^\ell_{jk}(x(t))\dot{x}^j(t)\dot{x}^k(t) = 0, \quad \ell = 1, \ldots, n \tag{4.60}$$

with

$$\Gamma^\ell_{jk} = \frac{1}{2} g^{\ell i}(g_{ji,k} + g_{ik,j} - g_{kj,i}). \tag{4.61}$$

The expressions Γ^ℓ_{jk} are called *Christoffel symbols*, after Elwin Bruno Christoffel, and so are the quantities obtained by pulling down the upper index,

$$\Gamma_{ijk} = \frac{1}{2}(g_{ji,k} + g_{ik,j} - g_{kj,i}). \tag{4.62}$$

Later on, we shall need

Lemma 4.1 *For any $p \in M$ and any tangent vector V at p, there exists a unique geodesic $c : [0, \epsilon) \to M$ for some $\epsilon > 0$, that is, a solution of (4.60), with*

$$c(0) = p, \qquad \dot{c}(0) = v. \tag{4.63}$$

---

[25] For a detailed treatment of the calculus of variations in general and the geodesic equations in particular, see for instance (Jost and Li-Jost 1998).

This follows from standard results about solutions of systems of second order ordinary differential equations (the Picard-Lindelöf Theorem) such as (4.60).[26].

For consistency, we also compute

$$\begin{aligned}\frac{d}{dt}(g_{ij}(x(t))\dot{x}^i\dot{x}^j) &= g_{ij}\ddot{x}^i\dot{x}^j + g_{ij}\dot{x}^i\ddot{x}^j + g_{ij,k}\dot{x}^i\dot{x}^j\dot{x}^k \\ &= -(g_{jk,\ell} + g_{\ell j,k} - g_{\ell k,j})\dot{x}^\ell\dot{x}^k\dot{x}^j + g_{\ell j,k}\dot{x}^k\dot{x}^\ell\dot{x}^j \text{ by (4.60)} \\ &= 0\end{aligned}$$

and therefore

$$g_{ij}(x(t))\dot{x}^i(t)\dot{x}^j(t) \equiv \text{const}, \tag{4.64}$$

that is, (4.58) holds. (4.64) simply means that the length of the tangent vector of a geodesic is constant.

4.4.5 Normal Coordinates

Riemann then used the freedom of the choice of coordinates to introduce special coordinates around a given point p. These coordinates are analogous to the polar coordinates in Euclidean space [see (4.7)]. In those Euclidean polar coordinates, the position of a point A is described by its distance t from the origin 0 and the direction ξ of the straight line from the origin 0 to A. That is, we center everything at 0 and then describe any other point by how far and in which direction we have to go from 0 to reach that point. The direction ξ can be described by a vector with components $\xi^1, \ldots, \xi^n$ with

$$\sum_{i=1}^{n} \xi^i\xi^i = 1, \tag{4.65}$$

that is, a direction is determined by a point on the unit sphere. Thus, we can represent A by a direction ξ and its distance t from the origin. (At the origin itself, that is, for $A = 0$, the direction ξ is undetermined, but that will cause no problems in the sequel.)

Riemann then transfers this scheme to a Riemannian manifold M. Let $p \in M$. A preliminary step consists in choosing coordinates for which p corresponds to the origin $0 \in \mathbb{R}^n$ and for which the metric tensor at p, that is, in the coordinates at 0 is given by the identity matrix,

$$g_{ij}(0) = \delta_{ij} := \begin{cases} 1 & \text{if } i = j \\ 0 & \text{else.} \end{cases} \tag{4.66}$$

[26] See for instance Jost (2005).

This can be achieved by a linear transformation that diagonalizes the metric tensor at 0.

Anyway, the metric structure in the tangent space T_pM when identified with $\mathbb{R}^n$ through our coordinates in which (4.66) holds then is the Euclidean one. We therefore choose the unit directions $\xi = (\xi^1, \dots, \xi^n)$ satisfying (4.65). For each such ξ, we consider the geodesic curve $c_\xi(s)$ that starts at 0 with the direction given by ξ, that is,

$$c_\xi(0) = p, \qquad \dot{c}_\xi(0) = \xi. \tag{4.67}$$

According to Lemma 4.1, such a geodesic $c_\xi(s)$ exists for sufficiently small $s > 0$. And conversely, every point $q \in M$ that is sufficiently close to p can be reached from p by a unique such short geodesic, that is, $q = c_\xi(t)$ for some ξ with (4.65) and some small $t \geq 0$. Therefore, we can characterize such a point q by a direction ξ and its distance $t \geq 0$ from p. Thus, we can use ξ and t as the coordinates for q. These are the Riemann normal coordinates that (after solving some minor technical problems over which Riemann glosses, but which are not hard) can be introduced in the vicinity of any point $p \in M$. p is a distinguished point for this particular coordinate system (of course, we can introduce such coordinates around any point in our manifold), because it corresponds to $0 \in \mathbb{R}^n$ in our coordinates, and at p, (4.66) holds. That is, the coordinates of a point in the vicinity of p are given by

$$x^i = t\xi^i, \quad i = 1, \dots, d. \tag{4.68}$$

For the understanding of the sequel, it will now be crucial to careful distinguish what will hold at 0 (that is, at p) only and what will be valid across the entire coordinate chart that we are going to construct. For instance, unless the metric is Euclidean, we can achieve (4.66) only at a single point p, here identified with the origin 0, but not at all other points of the chart.

The crucial point is that in those coordinates, any geodesic starting from 0 becomes a straight line, that is,

$$c_\xi(s) = s\xi. \tag{4.69}$$

With (4.66), (4.64) then yields

$$g_{ij}(x)x^i x^j = g_{ij}(0)x^i x^j = \sum_i x^i x^i. \tag{4.70}$$

Again, we point out that, unless the metric is Euclidean, formulas like (4.69), (4.70) hold only for geodesics starting at 0, but not necessarily for other geodesics connecting different points in our coordinate chart.

Equation (4.69) implies

$$\dot{c}_\xi(s) = \xi, \quad \ddot{c}_\xi(s) = 0, \tag{4.71}$$

and hence the geodesic Eq. (4.60) implies

$$\Gamma^\ell_{jk}(t\xi)\xi^j\xi^k = 0 \text{ for all } \ell = 1, \dots, n \tag{4.72}$$

and all ξ, t in our coordinate chart, or by pulling down the index i,

$$0 = \Gamma_{ijk}(t\xi)\xi^j\xi^k \text{ for all } i \text{ and all } \xi. \tag{4.73}$$

We first explore (4.73) for $t = 0$ to get

$$\Gamma_{ijk}(0) = 0 \text{ for all } i, j, k \tag{4.74}$$

because Γ_{ijk} is symmetric in the indices j, k. Thus

$$g_{ji,k}(0) + g_{ki,j}(0) - g_{jk,i}(0) = 0 \tag{4.75}$$

and adding the relation

$$g_{kj,i}(0) + g_{ij,k}(0) - g_{ki,j}(0) = 0 \tag{4.76}$$

which follows by a cyclic permutation of the indices, we get

$$g_{ij,k}(0) = 0 \text{ for all indices,} \tag{4.77}$$

and hence of course also

$$\Gamma^i_{jk}(0) = 0 \text{ for all } i, j, k \tag{4.78}$$

which, however, already follows from (4.74) or directly from (4.72).

Thus, at the center 0 of our coordinates, not only is the metric tensor diagonal (4.66), but also all its first derivatives vanish. In particular, since therefore we can make the first derivatives of the metric tensor disappear at any given point by a suitable choice of coordinates, these first derivatives cannot contain any geometric invariants that can distinguish a Riemannian metric from a Euclidean one.

We now explore (4.73) at other points in our coordinate chart, in order to eventually arrive at invariants in terms of the second derivatives of the metric tensor at 0. To simplify our notation, we shall usually omit the argument from our expressions, that is write for instance g_{ij} instead of $g_{ij}(x)$ or $g_{ij}(t\xi)$.

Further properties of Riemannian normal coordinates may best be seen by using *polar coordinates*, instead of the Euclidean ones. We therefore introduce on $\mathbb{R}^n$ the standard polar coordinates

$$(r, \varphi^1, \ldots, \varphi^{n-1}),$$

where $\varphi = (\varphi^1, \ldots, \varphi^{n-1})$ parametrizes the unit sphere S^{n-1}. In the two-dimensional case, we have introduced polar coordinates already in (4.7), and the general case is not substantially different. The precise formula for φ will be irrelevant for our purposes, and we simply view them as a nonredundant version of the directions $\xi^1, \ldots, \xi^n$ that needed to satisfy the constraint $\sum(\xi^i)^2 = 1$, in order to reduce their number of degrees of freedom from n to $n-1$, so that together with the radial variable r, we have the correct total number n of degrees of freedom. We express the metric in polar coordinates and write g_{rr} instead of g_{11}, because of the special role of r. We also write $g_{r\varphi}$ instead of $g_{1\ell}$, $\ell \in \{2, \dots, n\}$, and $g_{\varphi\varphi}$ as abbreviation for $(g_{k\ell})_{k,\ell=2,\dots,n}$. In particular, in these coordinates at 0 (which corresponds to the chosen point $p \in M$)

$$g_{rr}(0) = 1, \, g_{r\varphi}(0) = 0 \tag{4.79}$$

by (4.66) and since this holds for Euclidean polar coordinates.

We now utilize the geodesic Eq. (4.60) in a similar manner as before. In polar coordinates, the geometry of this equation becomes more transparent. In fact, the radial lines $\varphi \equiv$ const. are geodesic, and their formula (4.68) now becomes $x(t) = (t, \varphi_0)$, φ_0 fixed, and from (4.72), writing Γ^i_{rr} instead of Γ^i_{11} (that is, using r now also as an index),

$$\Gamma^i_{rr} = 0 \qquad \text{for all} \qquad i,$$

which means

$$g^{ik}(2g_{rk,r} - g_{rr,k}) = 0, \;\; \text{for all } i,$$

thus

$$2g_{rk,r} - g_{rr,k} = 0, \;\; \text{for all } k. \tag{4.80}$$

From this, we first obtain for $k = r$

$$g_{rr,r} = 0,$$

and with (4.79) then

$$g_{rr} \equiv 1. \tag{4.81}$$

Inserting this now into (4.80), we get

$$g_{r\varphi,r} = 0,$$

and then again with (4.79)

$$g_{r\varphi} \equiv 0. \tag{4.82}$$

Therefore, we can write the metric in polar coordinates in the abbreviated form

$$dr^2 + g_{\varphi\varphi}(r, \varphi)d\varphi^2. \tag{4.83}$$

with

$$g_{\varphi\varphi}(0, \varphi) = 0 \text{ and } g_{\varphi\varphi,r}(0, \varphi) = 0, \tag{4.84}$$

by (4.77).

After this general observation, which will be useful in Sect. 4.4.10 below, we now embark upon a computation that will reveal important identities satisfied by the *second* derivatives of the metric tensor. Putting

$$y_i = g_{ij}x^j, \tag{4.85}$$

we get

$$g_{ij,k}x^j = \frac{\partial y_i}{\partial x^k} - g_{ik}, \tag{4.86}$$

hence with (4.73)

$$\begin{aligned}
0 &= \frac{\partial y_i}{\partial x^k}x^k - g_{ik}x^k - \frac{1}{2}\left(\frac{\partial y_k}{\partial x^i}x^k - g_{ki}x^k\right)\\
&= \frac{\partial y_i}{\partial x^k}x^k - \frac{1}{2}\left(\frac{\partial y_k}{\partial x^i}x^k + y_i\right) \text{ using (4.85)}\\
&= \frac{\partial y_i}{\partial x^k}x^k - \frac{1}{2}\frac{\partial y_k x^k}{\partial x^i}\\
&= \frac{\partial y_i}{\partial x^k}x^k - x^i \text{ since } y_k x^k = g_{ik}x^i x^k = \sum x^i x^i \text{ by (4.70)}\\
&= \frac{\partial (y_i - x^i)x^k}{\partial x^k}.
\end{aligned}$$

If now $x^i = t\xi^i$ by (4.68), this yields

$$\frac{d}{dt}(y_i - x^i) = 0,$$

and since $y_i - x^i = 0$ for $t = 0$, it has to vanish for all t, and we get

$$y_i = g_{ik}x^k = x^i. \tag{4.87}$$

This implies in turn

$$\frac{\partial g_{ik}}{\partial x^j}x^k = \delta_{ij} - g_{ij}, \tag{4.88}$$

and since the right-hand side is symmetric in i and j, so then is the left hand side, and we obtain

$$\frac{\partial g_{ik}}{\partial x^j}x^k = \frac{\partial g_{jk}}{\partial x^i}x^k. \tag{4.89}$$

When we use this in (4.73), we can split that relation into two equations, and divide by t^2 to get

$$\frac{\partial g_{ik}(t\xi)}{\partial x^j}\xi^j\xi^k = 0 \quad \text{and} \tag{4.90}$$

$$\frac{\partial g_{jk}(t\xi)}{\partial x^i}\xi^j\xi^k = 0. \tag{4.91}$$

We now consider the second derivatives of the metric tensor at the origin 0.

$$g_{ij,k\ell} := \frac{\partial^2 g_{ij}}{\partial x^k \partial x^\ell}(0). \tag{4.92}$$

This expression is symmetric in i and j (because g_{ij} is symmetric), and in k and ℓ (because the derivatives with respect to x^k and x^ℓ commute), and we shall use this freely in the sequel. We differentiate (4.90) with respect to t and get at $t = 0$

$$g_{ik,j\ell}\xi^j \xi^k \xi^\ell = 0 \tag{4.93}$$

and by differentiating this with respect to ξ^m, we obtain

$$g_{im,k\ell}\xi^k \xi^\ell + g_{ij,m\ell}\xi^j \xi^\ell + g_{ij,km}\xi^j \xi^k = 0, \tag{4.94}$$

that is, using symmetries,

$$g_{im,k\ell} + g_{ik,\ell m} + g_{i\ell,mk} = 0 \tag{4.95}$$

for all indices i, m, k, ℓ. Similarly, from (4.91) we obtain

$$g_{im,k\ell} + g_{mk,i\ell} + g_{ki,m\ell} = 0. \tag{4.96}$$

Equations (4.95) and (4.96) then also yield the symmetry

$$g_{im,k\ell} = g_{k\ell,im}. \tag{4.97}$$

We now Taylor expand g_{ij} at 0 to second order.[27] Following Weyl, we write here δx instead of x, as Riemann does, for the variable with respect to which we expand near 0, because subsequently, dx and δx will both be interpreted geometrically as infinitesimal vectors. With the formulae (4.66), (4.77) and the definition (4.92), we thus have

$$g_{ij}(\delta x) = \delta_{ij} + \frac{1}{2} g_{ij,k\ell}\delta x^k \delta x^\ell + o\left(|\delta x|^2\right) \text{ for small } |\delta x|. \tag{4.98}$$

We thus have the expansion of the metric tensor

$$g_{ij}(\delta x) dx^i dx^j = \delta_{ij} dx^i dx^j + \frac{1}{2} g_{ij,k\ell}\delta x^k \delta x^\ell dx^i dx^j + o\left(|\delta x|^2\right). \tag{4.99}$$

[27] Taylor expansion means that we write a function f that is defined in some neighborhood of $0 \in \mathbb{R}^n$ and that is sufficiently often differentiable as

$$f(\delta x) = f(0) + \frac{\partial f}{\partial x^m}(0)\delta x^m + \frac{1}{2}\frac{\partial^2 f}{\partial x^k \partial x^\ell}(0)\delta x^k \delta x^\ell + o\left(|\delta x|^2\right)$$

where $o(|\delta x|^2)$ stands for a quantity with $\lim_{\delta x \to 0} \frac{o(|\delta x|^2}{|\delta x|^2} = 0$, but is otherwise unspecified.

We now put

$$R_{jik\ell} := \frac{1}{2}\big(g_{j\ell,ik} + g_{ik,j\ell} - g_{jk,i\ell} - g_{i\ell,jk}\big). \tag{4.100}$$

(We shall subsquently identify the $R_{ijk\ell}$ as the components of the Riemann curvature tensor.) We observe that these quantities possess the following symmetries

$$R_{ijk\ell} = -R_{jik\ell} \tag{4.101}$$

$$R_{ji\ell k} = -R_{jik\ell} \tag{4.102}$$

$$R_{jik\ell} + R_{ikj\ell} + R_{kji\ell} = 0 \tag{4.103}$$

$$R_{k\ell ji} = R_{jik\ell} \tag{4.104}$$

We now state

Theorem 4.1 *We can write (4.99) in the form*

$$g_{ij}(\delta x)dx^i dx^j = \delta_{ij}dx^i dx^j - \frac{1}{3}R_{jik\ell}\Delta x^{ij}\Delta x^{k\ell} + o(|\delta x|^2) \tag{4.105}$$

with

$$\Delta x^{rs} := \frac{1}{2}(\delta x^r dx^s - \delta x^s dx^r). \tag{4.106}$$

The Δx^{rs} are the components of the surface element representing the triangle with vertices $0, \delta x, dx$. The square of the area of that triangle is

$$\frac{1}{16}(g_{ik}g_{j\ell} - g_{i\ell}g_{jk})\Delta x^{ij}\Delta x^{k\ell} \tag{4.107}$$

The products $\Delta x^{ij}\Delta x^{k\ell}$ satisfy the same symmetries (4.101)–(4.104) as the $R_{jik\ell}$. This matches with the general considerations in Sect. 4.4.2, and if we instead we start with the expansion (4.105) and impose those symmetries on the coefficients $R_{jik\ell}$, then they are uniquely determined.

Proof By (4.95), (4.97), we can write

$$g_{ij,k\ell} = \frac{2}{3}g_{ij,k\ell} + \frac{1}{3}g_{ij,k\ell} = \frac{1}{3}(g_{ij,k\ell} + g_{k\ell,ij}) - \frac{1}{3}(g_{ik,\ell j} + g_{i\ell,jk}) \tag{4.108}$$

and so

$$
\begin{aligned}
g_{ij,k\ell}\delta x^k \delta x^\ell dx^i dx^j &= \left(\frac{1}{3}(g_{ij,k\ell}+g_{k\ell,ij}) - \frac{1}{3}(g_{jk,\ell i}+g_{i\ell,jk})\right)\delta x^k \delta x^\ell dx^i dx^j \\
&\quad \text{exchanging } i, j \text{ in the third term} \\
&= \left(\frac{1}{3}(g_{ik,j\ell}+g_{j\ell,ik}) - \frac{1}{3}(g_{kj,\ell i}+g_{i\ell,kj})\right) dx^i \delta x^j dx^k \delta x^\ell \\
&\quad \text{exchanging } j, k. \qquad (4.109)
\end{aligned}
$$

□

4.4.6 Riemann's Abstract Reasoning

Before we proceed with our tensor computations, and in order to put them into the general context, let us turn to the more abstract arguments that Riemann presents in his text. In (4.105), we have identified the second order term in the expansion (4.99) of the metric tensor. In contrast to the first derivatives of the metric tensor which can be made to vanish in appropriate coordinates, e.g., normal coordinates, and therefore do not carry any invariant geometric content, that second order term has an invariant geometric interpretation and therefore will not vanish for a general Riemannian metric. We shall also see below that for a two-dimensional manifold, i.e., a surface, in Euclidean 3-space, it coincides with the Gauss curvature of that surface, a geometric invariant identified by Gauss. For an n-dimensional manifold, it provides an invariant, later called the sectional curvature, to each surface direction at a point. Since the number of linearly independent two-dimensional subspaces of an n-dimensional vector space (taken as the tangent space at the point under consideration) is $\frac{n(n-1)}{2}$, we thus have found that many invariants attached to the metric tensor at each point of the manifold. And conversely, when we evaluate that curvature on $\frac{n(n-1)}{2}$ general, independent surface directions at a point, the entire curvature tensor at that point is determined. But this, according to Riemann's count, is precisely the right number of invariants that the metric tensor can contain. As already explained, the metric tensor g_{ij}, being symmetric, that is, $g_{ji} = g_{ij}$, has $\frac{n(n+1)}{2}$ independent components, but as we have the freedom of choosing the n coordinate functions, we can eliminate n degrees of freedom, and so, $\frac{n(n-1)}{2}$ remain. And as we have just argued, this is the number of degrees of freedom provided by the surface curvatures. In particular, the higher order terms ($o(|\delta x|^2)$) do not hide any further geometric invariants beyond those already contained in the curvature tensor. (We need to be somewhat careful here, however. A quadratic form $(h_{ij})_{i,j=1,\dots,n}$ is not determined when we evaluate it on n linearly independent vectors. For instance, the quadratic form in two dimensions

$$
(h_{ij}) = \begin{pmatrix} 1 & 0 \\ 0 & -1 \end{pmatrix}
$$

evaluated on a vector (v^1, v^2) via

$$v^i h_{ij} v^j$$

yields the value 0 for both vectors $(1, 1)$ and $(1, -1)$, without being equivalent to the form $\begin{pmatrix} 0 & 0 \\ 0 & 0 \end{pmatrix}$ itself. Something analogous occurs for curvature tensors as constructed in Sect. 4.4.7. We can take the product of a surface S_1 of constant curvature 1 and another surface S_{-1} of constant curvature -1 (see Sect. 4.4.10). The result is a 4-dimensional Riemannian manifold M. At each point $p \in M$, for all 6 surface directions spanned by two tangent vectors $V_1 \pm V_{-1}$, with V_α tangent to S_α, the curvature in that direction is 0. Still, the manifold is not flat.[28] It seems to me that Riemann was aware of such examples—certainly, he knew the theory of invariants of quadratic forms as developed by Gauss—and wanted to exclude them by his appeal to general situations and independent measurements at different points. The important point is that the directions taken at different points should not be systematically related.)

For a Euclidean metric, all these invariants have to vanish, as in that case, we have $g_{ij}(\delta x)dx^i dx^j = \delta_{ij} dx^i dx^j$ in (4.105). Riemann calls such a metric *flat*. Now, since the curvatures are geometric quantities and therefore independent of the coordinates chosen, they have to remain invariant under coordinate changes. Thus, when a metric is written in some coordinates and we want to check whether we can bring it into the Euclidean form $g_{ij}(x) = \delta_{ij}$, that is, if we want to find out whether our metric is simply the Euclidean metric in disguise, it is necessary and sufficient to check whether all its sectional curvatures, or equivalently, whether its sectional curvatures in $\frac{n(n-1)}{2}$ general, independent surface directions vanish. More generally, when we want to check whether two metrics $g_{ij}(x)dx^i dx^j$ and $h_{k\ell}(y)dy^k dy^\ell$ in different coordinates are the same, we simply check whether their curvatures are the same.

In somewhat more abstract terms, for finding out whether there exists a coordinate transformation $x = x(y)$ so that $g_{ij}(x)dx^i dx^j = g_{ij}(x(y))\frac{\partial x^i}{\partial y^k}\frac{\partial x^j}{\partial y^\ell}dy^k dy^\ell = h_{k\ell}(y)dy^k dy^\ell$, that is, whether one tensor can be transformed into the other, it is necessary and sufficient to check whether another tensor, the curvature tensor, is the same for the two metrics. In formulae, when we denote the curvature tensor in the y-coordinates by $S_{k\ell rs}$, the condition is

$$R_{ijpq}(x)dx^i dx^j dx^p dx^q = S_{k\ell rs}(x(y))dy^k dy^\ell dy^r dy^s. \tag{4.110}$$

Of course, it needs to be verified that this R_{ijpq} possesses the correct transformation behavior for a tensor, but from abstract principles this is plausible, if not evident.

Riemann's parameter counting argument, on which the preceding reasoning depends, is somewhat heuristic, and so far, nobody has really made it mathematically precise. Therefore, the detailed mathematical reasoning of Riemann himself in his unpublished Paris Academy essay and of his successors (who did not know that essay) proceeded somewhat differently. One derives the necessary and sufficient conditions for a metric in local coordinates to be

[28] See Scala (2001), where it is asserted that Riemann's corresponding assertion was incorrect.

flat, or more generally, for two metrics to be the same, from an analysis of the transformation formula for a metric under coordinate changes and finds a tensor composed of certain second and first derivatives of the metric as containing those conditions. In normal coordinates, this tensor is seen to reduce to (4.100). In that way, the covariant form of the curvature tensor is found, and it is shown, indeed, to provide the necessary and sufficient conditions for the equivalence of two metrics. We shall now turn to that mathematical reasoning.

4.4.7 Flatness and Curvature

In this section, we want to verify Riemann's claim that the curvature determines the metric by a computation. Riemann himself carried out such a computation in his Paris Academy essay for the case of a flat metric. More precisely, he showed that a metric $g_{ij}(x)dx^i dx^j$ can be transformed into the flat metric $\delta_{k\ell}dy^k dy^\ell$ by some coordinate transformation $y = y(x)$ if and only if its curvature tensor vanishes. We shall first present here the reasoning in Riemann's Paris essay, which replaces the abstract arguments of the habilitation address by concrete tensor type computations. Although some of the conventions and rules of tensor calculus were established only by Riemann's successors, we shall utilize them here. This will not affect the essence of the argument. Likewise, some of the details will be arranged somewhat differently than Riemann did.

Given a Riemannian metric $g_{ij}(x)dx^i dx^j$ in the coordinates $x^1, \ldots, x^n$, we ask whether there exists a coordinate transformation $x = x(y)$ so that in the coordinates $y^1, \ldots, y^n$, the metric has the Euclidean form $\sum_k dy^k dy^k$ which we can also write as $\delta_{k\ell}dy^k dy^\ell$ with a Kronecker symbol $\delta_{k\ell}$ to avail ourselves of the summation convention of tensor calculus. We cannot directly read this off from the coefficients $g_{ij}(x)$. We rather need to derive differential conditions and check whether we can solve them. The question is whether there exists a solution to the transformation

$$g_{pq}(x(y))\frac{\partial x^p}{\partial y^k}\frac{\partial x^q}{\partial y^\ell}dy^k dy^\ell = \delta_{k\ell}dy^k dy^\ell. \tag{4.111}$$

Equating coefficients in (4.111) and multiplying by $\frac{\partial y^\ell}{\partial x^j}$ yields

$$g_{pq}(x(y))\frac{\partial x^p}{\partial y^k}\frac{\partial x^q}{\partial y^\ell}\frac{\partial y^\ell}{\partial x^j} = \delta_{k\ell}\frac{\partial y^\ell}{\partial x^j},$$

that is,

$$g_{pj}(x(y))\frac{\partial x^p}{\partial y^k} = \frac{\partial y^k}{\partial x^j}, \tag{4.112}$$

and multiplying by $\frac{\partial y^k}{\partial x^i}$ then yields

$$g_{ij} = \sum_k \frac{\partial y^k}{\partial x^i} \frac{\partial y^k}{\partial x^j}. \tag{4.113}$$

Alternatively, this formula could have been derived also from the analogue of (4.111) for the inverse transformation $y = y(x)$, that is,

$$g_{ij} dx^i dx^j = \delta_{k\ell} \frac{\partial y^k}{\partial x^i} \frac{\partial y^\ell}{\partial x^j} dy^k dy^\ell. \tag{4.114}$$

From (4.112), we also obtain by multiplication with g^{mj}

$$\frac{\partial x^m}{\partial y^k} = g^{mj} \frac{\partial y^k}{\partial x^j}$$

and by further multiplication with $\frac{\partial x^i}{\partial y^k}$ and renaming the index m to j then

$$g^{ij} = \delta^{k\ell} \frac{\partial x^i}{\partial y^\ell} \frac{\partial x^j}{\partial y^k}, \tag{4.115}$$

and from (4.115) in turn by the same procedure

$$g^{ij} \frac{\partial y^\ell}{\partial x^i} \frac{\partial y^k}{\partial x^j} = \delta^{k\ell}. \tag{4.116}$$

Upon some reflection, the relations that we have obtained so far are more or less obvious, and in fact, constitute easy applications of tensor calculus, but we now need to interpret and solve them. (4.113) is a system of differential equations for the coordinates $y^1, \ldots, y^n$ as functions of the $x^1, \ldots, x^n$. In order to solve it, we differentiate it with respect to x^m to get

$$\sum_k \frac{\partial^2 y^k}{\partial x^i \partial x^m} \frac{\partial y^k}{\partial x^j} + \sum_k \frac{\partial^2 y^k}{\partial x^j \partial x^m} \frac{\partial y^k}{\partial x^i} = g_{ij,m}. \tag{4.117}$$

Since the same equations obtain for $g_{im,j}, g_{mj,i}$, we obtain

$$\sum_k \frac{\partial^2 y^k}{\partial x^i \partial x^j} \frac{\partial y^k}{\partial x^m} = \frac{1}{2} \left(g_{im,j} + g_{jm,i} - g_{ij,m} \right). \tag{4.118}$$

We multiply this by $\frac{\partial y^\ell}{\partial x^p} g^{pm}$ and use (4.116) to obtain

$$\frac{\partial^2 y^\ell}{\partial x^i \partial x^j} = \frac{1}{2} g^{pm} \left(g_{im,j} + g_{jm,i} - g_{ij,m} \right) \frac{\partial y^\ell}{\partial x^p} = \Gamma^p_{ij} \frac{\partial y^\ell}{\partial x^p}. \tag{4.119}$$

As a side remark that will be needed later, we point out that with the computational scheme established, we can also compute the general transformation formula for the Christoffel symbols of an arbitrary metric under a change of coordinates $y = y(x)$. Denoting the Christoffel symbols for the y-coordinates by $H^q_{\ell m}$, the result is

$$\Gamma^k_{ij} = \frac{\partial y^\ell}{\partial x^i}\frac{\partial y^m}{\partial x^j}\frac{\partial x^k}{\partial y^q}H^q_{\ell m} + \frac{\partial^2 y^q}{\partial x^i \partial x^j}\frac{\partial x^k}{\partial y^q}, \tag{4.120}$$

or, equivalently, in the other direction,

$$H^q_{\ell m} = \frac{\partial x^i}{\partial y^\ell}\frac{\partial x^j}{\partial y^m}\frac{\partial y^q}{\partial x^k}\Gamma^k_{ij} + \frac{\partial^2 x^k}{\partial y^m \partial y^\ell}\frac{\partial y^q}{\partial x^k}. \tag{4.121}$$

This formula shows that the Christoffel symbols *do not transform as tensors*, because the transformation rule involves also the second derivatives of the coordinate change.

We now want to check under which conditions the Eq. (4.119) can be solved. Necessary conditions come from the commutativity of derivatives. First, we see that the condition $\frac{\partial^2 y^\ell}{\partial x^i \partial x^j} = \frac{\partial^2 y^\ell}{\partial x^j \partial x^i}$ is satisfied because the Christoffel symbols are symmetric, $\Gamma^p_{ij} = \Gamma^p_{ji}$. Now we test for the commutativity of third derivatives, $\frac{\partial^3 y^\ell}{\partial x^i \partial x^j \partial x^k} = \frac{\partial^3 y^\ell}{\partial x^i \partial x^k \partial x^j}$. Thus, we need to differentiate the right hand side of (4.119) with respect to x^k and then take the analogous equations for $\frac{\partial^2 y^\ell}{\partial x^i \partial x^k}$ and differentiate them with respect to x^j. The results should then agree, which leads us to the condition

$$\begin{aligned} 0 &= \left(\frac{\partial \Gamma^p_{ij}}{\partial x^k} - \frac{\partial \Gamma^p_{ik}}{\partial x^j}\right)\frac{\partial y^\ell}{\partial x^p} + \Gamma^p_{ij}\frac{\partial^2 y^\ell}{\partial x^p \partial x^k} - \Gamma^p_{ik}\frac{\partial^2 y^\ell}{\partial x^p \partial x^j} \\ &= \left(\frac{\partial \Gamma^q_{ij}}{\partial x^k} - \frac{\partial \Gamma^q_{ik}}{\partial x^j}\right)\frac{\partial y^\ell}{\partial x^q} + \left(\Gamma^p_{ij}\Gamma^q_{pk} - \Gamma^p_{ik}\Gamma^q_{pj}\right)\frac{\partial y^\ell}{\partial x^q} \end{aligned} \tag{4.122}$$

by using (4.119) once more in the last step and renaming an index from p to q. These relations have to hold for all ℓ, and since for each q, there is at least one ℓ with $\frac{\partial y^\ell}{\partial x^q} \neq 0$ as the transformation $y = y(x)$ is invertible, we conclude the necessary condition

$$\frac{\partial \Gamma^q_{ij}}{\partial x^k} - \frac{\partial \Gamma^q_{ik}}{\partial x^j} + \Gamma^p_{ij}\Gamma^q_{pk} - \Gamma^p_{ik}\Gamma^q_{pj} = 0 \text{ for all } i, j, k, q. \tag{4.123}$$

This is the central result. We have seen that this condition is necessary for solving (4.119) and hence for finding a transformation with (4.111), that is, for transforming the given Riemannian metric $g_{ij}(x)$ into a flat metric. In fact, the conditions (4.123) are also sufficient for this in the following sense. In general, one can solve a system of equations of the form

$$\frac{\partial g}{\partial x^j} = \phi_j \tag{4.124}$$

for given functions ϕ_j and an unknown function g in a simply connected domain if and only if the conditions for the commutativity of the second partial derivatives of g, $\frac{\partial^2 g}{\partial x^j \partial x^k} = \frac{\partial^2 g}{\partial x^k \partial x^j}$, that is,

$$\frac{\partial \phi_j}{\partial x^k} = \frac{\partial \phi_k}{\partial x^j} \text{ for all } j, k \tag{4.125}$$

are satisfied. This general result is known as the Theorem of Frobenius, but obviously, some version of it was already known to Riemann and his contemporaries.[29] And in fact, (4.123) precisely provides the integrability conditions (4.125) for the differential equations (4.119) for the functions $\frac{\partial y^\ell}{\partial x^i}$. Thus, (4.123) is the necessary and sufficient local condition for $g_{ij}(x)$ to be Euclidean, indeed.

Moreover, if we define

Definition 4.2 *The Riemann curvature components are*

$$\begin{aligned} R^q_{ikj} &:= \frac{\partial \Gamma^q_{ij}}{\partial x^k} - \frac{\partial \Gamma^q_{ik}}{\partial x^j} + \Gamma^p_{ij}\Gamma^q_{pk} - \Gamma^p_{ik}\Gamma^q_{pj} \\ \text{and } R_{\ell ijk} &:= g_{\ell q} R^q_{ijk}, \end{aligned} \tag{4.126}$$

then in normal coordinates, we have

$$R_{\ell ijk} = \frac{1}{2}\big(g_{k\ell,ij} + g_{ij,k\ell} - g_{ik,j\ell} - g_{j\ell,ik}\big), \tag{4.127}$$

which is the same as (4.100). Thus, if we can verify that (4.126) defines a tensor, the Riemann curvature tensor

$$R_{\ell ijk} dx^\ell dx^i dx^j dx^k, \tag{4.128}$$

then everything fits together. That is, (4.126) will give the formula for the curvature tensor (4.105) in arbitrary coordinates, and in turn, Theorem 4.1 and the discussion around provide us with a geometric explanation of curvature. In order to see that (4.126) defines a tensor, we need to check that it correctly transforms under coordinate changes. This can either be done directly from the formula given, but this is a rather cumbersome computation, or it can be derived from more general arguments. A natural possibility is to generalize the question that we have addressed, to obtain a criterion whether a given metric $g_{ij}dx^i dx^j$ is Euclidean, and ask for a criterion when two metrics

$$g_{ij}(x)dx^i dx^j \quad \text{and} \quad h_{rs}(y)dy^r dy^s \tag{4.129}$$

can be locally transformed into each other. This can, of course, be done along the lines just presented. In the course of the resulting computations, you will see that the necessary and

29 Frobenius (1877) proves a general theorem that puts several previous results into a systematic perspective.

sufficient condition is that their curvatures agree, and those computations will also uncover the tensorial transformation behavior of the curvature.[30]

Instead of proceeding in that manner, I shall now present the general computation within a more abstract framework. A reader who is interested only in the historical aspects may therefore wish to skip the rest of this section. Its purpose is to clarify the geometric content. The fundamental concept will be that of the covariant derivative. That concept was not yet present in Riemann's work, but only emerged from the work of Riemann's successors, that is, Lipschitz, Christoffel, Ricci, Levi-Civita and Weyl. In that sense, the approach is somewhat anachronistic, as not yet contained in Riemann's own work, but only representing a further development of his seminal ideas. However, one can also simply view it as a systematic and efficient way to organize the computations that are needed to show that curvature transforms as a tensor under coordinate changes and that two metrics are equivalent under a coordinate change if and only if their curvature tensors are the same. That latter fact is the key for Riemann's reasonings.

The concept of the covariant derivative of Levi-Civita that we are now going to introduce will provide a geometric interpretation to the Christoffel symbols.[31]

Definition 4.3 Let $V(x) = v(x)^i \frac{\partial}{\partial x^i}$ be a vector field, that is, a family of tangent vectors depending smoothly on the position x. The covariant derivative of V in the direction $\frac{\partial}{\partial x^j}$ then is

$$\nabla_{\frac{\partial}{\partial x^j}} V := \frac{\partial v^i}{\partial x^j} \frac{\partial}{\partial x^i} + v^i \Gamma^k_{ij} \frac{\partial}{\partial x^k}. \tag{4.130}$$

Eq. (4.130) is also called the Levi-Civita connection.

At this point, the reader should be warned that this is a definition of the Γ^k_{ij} that need not have anything to do or be compatible with our earlier definition (4.61). But we shall see in a moment that the Γ^k_{ij} as defined in (4.130) transform by the same rule (4.119) as those defined in (4.61) did. Therefore, we can indeed use our old Christoffel symbols in the Definition (4.130). To see this, from (4.130), we have

$$\nabla_{\frac{\partial}{\partial x^j}} \frac{\partial}{\partial x^i} = \Gamma^k_{ij} \frac{\partial}{\partial x^k}. \tag{4.131}$$

---

[30] These computations were first provided by Christoffel (1869) and Lipschitz (1869) in two articles in the same journal issue.

[31] The concept of such a derivative was already essentially contained in the work of Christoffel. It is named after Levi-Civita because the latter then developed a notion of parallel transport from such a covariant derivative.

For new coordinates $y = y(x)$, we denote their Christoffel symbols by $H^q_{\ell m}$, and they thus satisfy

$$\nabla_{\frac{\partial}{\partial y^\ell}} \frac{\partial}{\partial y^m} = H^q_{\ell m} \frac{\partial}{\partial y^q}. \tag{4.132}$$

On the other hand, rewriting (4.132) in the old coordinates and writing $\frac{\partial}{\partial x^j} = \frac{\partial y^q}{\partial x^j} \frac{\partial}{\partial y^q}$ etc. yields

$$\begin{aligned} H^q_{\ell m} \frac{\partial}{\partial y^q} &= \nabla_{\frac{\partial x^i}{\partial y^\ell} \frac{\partial}{\partial x^i}} \left(\frac{\partial x^j}{\partial y^m} \frac{\partial}{\partial x^j} \right) \\ &= \frac{\partial x^i}{\partial y^\ell} \frac{\partial x^j}{\partial y^m} \frac{\partial y^q}{\partial x^k} \Gamma^k_{ij} \frac{\partial}{\partial y^q} + \frac{\partial^2 x^j}{\partial y^m \partial y^\ell} \frac{\partial y^q}{\partial x^j} \frac{\partial}{\partial y^q}, \end{aligned}$$

that is,

$$H^q_{\ell m} = \frac{\partial x^i}{\partial y^\ell} \frac{\partial x^j}{\partial y^m} \frac{\partial y^q}{\partial x^k} \Gamma^k_{ij} + \frac{\partial^2 x^j}{\partial y^m \partial y^\ell} \frac{\partial y^q}{\partial x^j}, \tag{4.133}$$

which is the same as (4.121), but obtained more easily.

When the Christoffel symbols vanish, the covariant derivative reduces to the ordinary derivative. In other words, the covariant derivative is a scheme of taking the derivative of a vector field in the context of a Riemannian metric, as we shall now explain in more detail.

First of all, instead of $\frac{\partial}{\partial x^j}$, we can use a general vector $W = w^j \frac{\partial}{\partial x^j}$ and put

$$\nabla_{w^j \frac{\partial}{\partial x^j}} V := w^j \frac{\partial v^i}{\partial x^j} \frac{\partial}{\partial x^i} + v^i w^j \Gamma^k_{ij} \frac{\partial}{\partial x^k}. \tag{4.134}$$

In this way, the derivative becomes tensorial with respect to the direction W in which it is taken. It is not tensorial with respect to V, however, as it involves taking derivatives of the coefficients of V. Rather, like any derivative, it satisfies a Leibniz product rule

$$\nabla_{\frac{\partial}{\partial x^j}} f(x) V(x) = \frac{\partial f(x)}{\partial x^j} V(x) + f(x) \nabla_{\frac{\partial}{\partial x^j}} V(x) \tag{4.135}$$

when $f(x)$ is a differentiable function. Of course, the covariant derivative as such transforms correctly. That means that when we change coordinates from x to y and correspondingly transform the vectors W and V, then the covariant derivative $\nabla_W V$ is transformed into the covariant derivative of the transformed vectors, constructed with the Christoffel symbols in the new coordinates. A consequence is that the Christoffel symbols themselves then cannot transform as tensors. Their transformation rule (4.133) also involves derivatives of the coordinate change. This fits with the results of Sect. 4.4.5 where we have seen that we can always transform them to 0 at a given point, and that they therefore cannot encode geometric invariants.

Like the ordinary derivative, the covariant derivative (4.130) satisfies a commutation property:

$$\nabla_{\frac{\partial}{\partial x^j}} \frac{\partial}{\partial x^i} = \Gamma_{ij}^k \frac{\partial}{\partial x^k} = \Gamma_{ji}^k \frac{\partial}{\partial x^k} = \nabla_{\frac{\partial}{\partial x^i}} \frac{\partial}{\partial x^j}, \tag{4.136}$$

because the Christoffel symbols are symmetric with respect to the lower indices. In fact, we also see that even if the Christoffel symbols were not symmetric and we could put

$$T_{ij} := \nabla_{\frac{\partial}{\partial x^j}} \frac{\partial}{\partial x^i} - \nabla_{\frac{\partial}{\partial x^i}} \frac{\partial}{\partial x^j}, \tag{4.137}$$

but this then would no longer vanish. Still, this does not yet define a tensor, because the derivatives of the coefficients do not yet drop out, even though we take the difference. But we can remedy this. For vector fields $W(x) = w^j(x)\frac{\partial}{\partial x^j}$, $Z(x) = z^k(x)\frac{\partial}{\partial x^k}$, we define the *Lie bracket*

$$[Z, W] := z^k \frac{\partial w^j}{\partial x^k} \frac{\partial}{\partial x^j} - w^j \frac{\partial z^k}{\partial x^j} \frac{\partial}{\partial x^k}. \tag{4.138}$$

We then put

$$\begin{aligned} T(Z, W) &:= \nabla_Z W - \nabla_W Z - [Z, W] \\ &= z^k w^j \nabla_{\frac{\partial}{\partial x^k}} \frac{\partial}{\partial x^j} + z^k \frac{\partial w^j}{\partial x^k} \frac{\partial}{\partial x^j} \\ &\quad - w^j z^k \nabla_{\frac{\partial}{\partial x^j}} \frac{\partial}{\partial x^k} - w^j \frac{\partial z^k}{\partial x^j} \frac{\partial}{\partial x^k} \\ &\quad - z^k \frac{\partial w^j}{\partial x^k} \frac{\partial}{\partial x^j} + w^j \frac{\partial z^k}{\partial x^j} \frac{\partial}{\partial x^k} \\ &= z^k w^j \left(\nabla_{\frac{\partial}{\partial x^k}} \frac{\partial}{\partial x^j} - \nabla_{\frac{\partial}{\partial x^j}} \frac{\partial}{\partial x^k} \right). \end{aligned} \tag{4.139}$$

Thus, by subtracting the Lie bracket from the difference of the covariant derivatives, we have made all derivatives of the coefficients disappear, and consequently, T defines a tensor.[32] This observation, that we can restore a tensorial transformation behavior by taking differences, will be important in a moment. Since $[\frac{\partial}{\partial x^k}, \frac{\partial}{\partial x^j}] = 0$, the coefficients of the tensor T from (4.139) are given by (4.137). Of course, to repeat, for our covariant derivative ∇, the torsion tensor T vanishes identically. We have introduced it here only in order to explain a general point that will become relevant for the curvature tensor R which in general does not vanish.

The reason behind the definition (4.130) involving the Christoffel symbols of a metric $g = (g_{ij})$ is that in that way, the covariant derivative satisfies a product rule for that metric [using the notation (4.41)]:

$$\frac{\partial}{\partial x^j} g(V, W) = g\left(\nabla_{\frac{\partial}{\partial x^j}} V, W\right) + g\left(V, \nabla_{\frac{\partial}{\partial x^j}} W\right), \tag{4.140}$$

---

[32] In the general theory of connections, as developed by Hermann Weyl and others, one defines an operation of the form (4.130) with abstract symbols Γ_{ij}^k that need not come from a metric and that need not be symmetric. The tensor T_{ij} then is called the torsion tensor of the connection.

as one readily checks. We can also interpret the relation (4.140) by saying that when we take a derivative of $g(V, W)$, then there are three objects involved, the vectors V, W and the metric g, but the rule is such that we only need to differentiate V and W, but not g, that is, the derivative of g vanishes. One also expresses this by saying that g is covariantly constant.

This also allows us to check the consistency with the earlier definition (4.61). Indeed, in local coordinates, inserting $V = \frac{\partial}{\partial x^i}$, $W = \frac{\partial}{\partial x^k}$, (4.140) says

$$g_{ik,j} = g_{\ell k}\Gamma^{\ell}_{ij} + g_{i\ell}\Gamma^{\ell}_{jk}, \tag{4.141}$$

which is equivalent to the definition (4.61) of the Christoffel symbols. (Conceptually, the issue is slightly different. We could have defined a covariant derivative (4.130) with arbitrary symbols Γ^k_{ij} that need not have anything to do with a metric nor even satisfy (4.136). Only when we relate the Christoffel symbols to the metric tensor via (4.61) do we get both relations (4.136) and (4.140). The theory of general such covariant derivatives that were introduced by Weyl, also called affine connections, while important in modern geometry and theoretical physics, however, goes beyond what is contained in Riemann's text, and will therefore not be further explored here.)[33]

We return to (4.131). This formula shows that the Christoffel symbols are the coefficients of the covariant derivative, or putting it the other way around, the Christoffel symbols define a notion of a derivative. We can interpret the non-tensorial transformation formula (4.133) as an equation that the second derivatives of a coordinate transformation have to satisfy,

$$\frac{\partial^2 x^j}{\partial y^m \partial y^\ell} = H^q_{\ell m}\frac{\partial x^j}{\partial y^q} - \frac{\partial x^i}{\partial y^\ell}\frac{\partial x^h}{\partial y^m}\Gamma^j_{ih}. \tag{4.142}$$

This equation will become important in a moment, and so, it might be useful to reflect a little about its conceptual status. Of course, as such, a coordinate transformation can be arbitrary, and will not satisfy any constraint. Here, however, we have a metric tensor g_{ij} with Christoffel symbols Γ^h_{ij}, and we have another one, $h_{k\ell}$ with Christoffel symbols $H^q_{k\ell}$, and we shall ask that the transformation $x = x(y)$ transform the latter into the former. That is, we ask that the two metrics g_{ij} and $h_{k\ell}$ be the same, just written in different coordinates. And in that case, the coordinate transformation needs to satisfy the constraint (4.142). Conversely, we can then ask under which conditions on the two collections of Christoffel symbols we can solve (4.142).

For this purpose, we start with the following observation. We have seen in (4.136) that first covariant derivatives commute. This no longer is the case for second derivatives. We set out to compute their difference. We put $\Gamma^k_{ij,\ell} := \frac{\partial \Gamma^k_{ij}}{\partial x^\ell}$. Then

[33] See, for instance, Jost (2017).

$$\nabla_{\frac{\partial}{\partial x^k}} \nabla_{\frac{\partial}{\partial x^j}} \left(v^i \frac{\partial}{\partial x^i} \right) = \frac{\partial^2 v^i}{\partial x^k \partial x^j} \frac{\partial}{\partial x^i} + v^i \Gamma^\ell_{km} \Gamma^m_{ij} \frac{\partial}{\partial x^\ell} + v^i \Gamma^m_{ij,k} \frac{\partial}{\partial x^m}$$
$$+ \Gamma^m_{ij} \frac{\partial v^i}{\partial x^k} \frac{\partial}{\partial x^m} + \Gamma^\ell_{ki} \frac{\partial v^i}{\partial x^j} \frac{\partial}{\partial x^\ell},$$

and hence

$$\nabla_{\frac{\partial}{\partial x^k}} \nabla_{\frac{\partial}{\partial x^j}} \left(v^i \frac{\partial}{\partial x^i} \right) - \nabla_{\frac{\partial}{\partial x^j}} \nabla_{\frac{\partial}{\partial x^k}} \left(v^i \frac{\partial}{\partial x^i} \right) = v^i \left(\Gamma^\ell_{ij,k} - \Gamma^\ell_{ik,j} + \Gamma^\ell_{km} \Gamma^m_{ij} - \Gamma^\ell_{jm} \Gamma^m_{ik} \right) \frac{\partial}{\partial x^\ell}$$
$$=: v^i R^\ell_{ikj} \frac{\partial}{\partial x^\ell}, \tag{4.143}$$

that is, defining the curvature coefficients R^ℓ_{ikj} as the commutators of second covariant derivatives. This is tensorial in V, and in order to obtain the tensorial transformation behavior also with respect to the other entries, we proceed as with the tensor T above.

Definition 4.4 The *Riemann curvature tensor* is

$$R(Z, W)V := \nabla_Z \nabla_W V - \nabla_W \nabla_Z V - \nabla_{[Z,W]} V. \tag{4.144}$$

and

$$R(Z, W, V, Y) := g(R(Z, W)V, Y) \tag{4.145}$$

with the Riemannian metric g.

We also write down the formulae in local indices once more, as in Definition 4.2,

$$R^\ell_{ikj} = \Gamma^\ell_{ij,k} - \Gamma^\ell_{ik,j} + \Gamma^\ell_{km} \Gamma^m_{ij} - \Gamma^\ell_{jm} \Gamma^m_{ik} \tag{4.146}$$
$$R_{mikj} = g_{m\ell} R^\ell_{ikj}. \tag{4.147}$$

In normal coordinates as constructed in Sect. 4.4.5, all the Christoffel symbols vanish at the point under consideration, and since the metric tensor is given by the identity matrix there, we also have

$$\Gamma^k_{ij,\ell} = \frac{1}{2}(g_{ik,j\ell} + g_{jk,i\ell} - g_{ij,k\ell}), \tag{4.148}$$

and we therefore obtain

$$R_{mikj} = \frac{1}{2}(g_{jm,ik} + g_{ik,jm} - g_{ij,km} - g_{km,ij}), \tag{4.149}$$

which agrees with (4.100). Thus, the $R_{jik\ell}$ appearing in the expansion (4.105) are the coefficients of the curvature tensor in normal coordinates. Thus, Theorem 4.1 provides us with a geometric interpretation of the curvature tensor.

We now return to (4.142) and ask under which conditions we can solve it. To recall, we have two Riemannian metrics $g_{ij}dx^i dx^j$ and $h_{k\ell}dy^k dy^\ell$ with their corresponding Christoffel symbols, and we ask when we can find a coordinate transformation $x = x(y)$ that transforms one into the other. We have identified (4.142) as a necessary condition. Equation (4.142) prescribes all the second derivatives of the coordinate transformation. Now, for any smooth function, its derivatives commute. We first observe that (4.142) implies that the second derivatives commute,

$$\frac{\partial^2 x^j}{\partial y^\ell \partial y^m} = \frac{\partial^2 x^j}{\partial y^m \partial y^\ell}; \tag{4.150}$$

indeed, this follows from the symmetry of the Christoffel symbols with respect to their lower indices, see (4.136). We now check under which conditions also third derivatives commute, that is, when we have

$$\frac{\partial^3 x^j}{\partial y^m \partial y^\ell \partial y^r} = \frac{\partial^3 x^j}{\partial y^m \partial y^r \partial y^\ell}. \tag{4.151}$$

For that purpose, we differentiate (4.142) and get

$$\begin{aligned}
\frac{\partial^3 x^j}{\partial y^m \partial y^\ell \partial y^r} &= H^q_{\ell m,r}\frac{\partial x^j}{\partial y^q} + H^q_{\ell m}\frac{\partial^2 x^j}{\partial y^q \partial y^r} \\
&\quad - \frac{\partial x^i}{\partial y^\ell}\frac{\partial x^h}{\partial y^m}\Gamma^j_{ih,f}\frac{\partial x^f}{\partial y^r} - \frac{\partial^2 x^i}{\partial y^\ell \partial y^r}\frac{\partial x^h}{\partial y^m}\Gamma^j_{ih} - \frac{\partial x^i}{\partial y^\ell}\frac{\partial^2 x^h}{\partial y^m \partial y^r}\Gamma^j_{ih} \\
&= H^q_{\ell m,r}\frac{\partial x^j}{\partial y^q} + H^q_{\ell m}H^p_{qr}\frac{\partial x^j}{\partial y^p} \\
&\quad - H^q_{\ell m}\frac{\partial x^i}{\partial y^q}\frac{\partial x^h}{\partial y^r}\Gamma^j_{ih} \\
&\quad - \frac{\partial x^i}{\partial y^\ell}\frac{\partial x^h}{\partial y^m}\Gamma^j_{ih,f}\frac{\partial x^f}{\partial y^r} \\
&\quad - H^q_{\ell r}\frac{\partial x^i}{\partial y^q}\frac{\partial x^k}{\partial y^m}\Gamma^j_{ih} - H^q_{rm}\frac{\partial x^h}{\partial y^q}\frac{\partial x^i}{\partial y^\ell}\Gamma^j_{ih} \\
&\quad + \frac{\partial x^h}{\partial y^m}\Gamma^j_{ih}\frac{\partial x^f}{\partial y^\ell}\frac{\partial x^h}{\partial y^r}\Gamma^i_{fh} + \frac{\partial x^i}{\partial y^\ell}\Gamma^j_{ih}\frac{\partial x^k}{\partial y^r}\frac{\partial x^f}{\partial y^m}\Gamma^h_{kf},
\end{aligned}$$

where we have inserted the Eq. (4.142) for the second derivatives. Likewise,

$$\begin{aligned}\frac{\partial^3 x^j}{\partial y^m \partial y^r \partial y^\ell} &= H^q_{rm,\ell}\frac{\partial x^j}{\partial y^q} + H^q_{rm}H^p_{q\ell}\frac{\partial x^j}{\partial y^p} \\ &\quad - H^q_{rm}\frac{\partial x^i}{\partial y^q}\frac{\partial x^h}{\partial y^\ell}\Gamma^j_{ih} \\ &\quad - \frac{\partial x^i}{\partial y^r}\frac{\partial x^h}{\partial y^m}\Gamma^j_{ih,f}\frac{\partial x^f}{\partial y^\ell} \\ &\quad - H^q_{r\ell}\frac{\partial x^i}{\partial y^q}\frac{\partial x^k}{\partial y^m}\Gamma^j_{ih} - H^q_{\ell m}\frac{\partial x^h}{\partial y^q}\frac{\partial x^i}{\partial y^r}\Gamma^j_{ih} \\ &\quad + \frac{\partial x^h}{\partial y^m}\Gamma^j_{ih}\frac{\partial x^f}{\partial y^r}\frac{\partial x^h}{\partial y^\ell}\Gamma^i_{fh} + \frac{\partial x^i}{\partial y^r}\Gamma^j_{ih}\frac{\partial x^k}{\partial y^\ell}\frac{\partial x^f}{\partial y^m}\Gamma^h_{kf},\end{aligned}$$

We observe that the mixed terms, with products of the Christoffel symbols Γ and H, are the same in both expressions, again because of the symmetry $\Gamma^j_{ih} = \Gamma^j_{hi}$. Therefore, they will drop out when we take the difference. For the same reason, the first terms in the last lines agree. Renaming some summation indices, we therefore can write the difference as

$$\begin{aligned}&\frac{\partial^3 x^j}{\partial y^m \partial y^\ell \partial y^r} - \frac{\partial^3 x^j}{\partial y^m \partial y^r \partial y^\ell} \\ &= H^q_{\ell m,r}\frac{\partial x^j}{\partial y^q} - H^q_{rm,\ell}\frac{\partial x^j}{\partial y^q} + H^p_{rm}H^q_{p\ell}\frac{\partial x^j}{\partial y^q} - H^p_{rm}H^q_{p\ell}\frac{\partial x^j}{\partial y^q} \\ &\quad -\left(\Gamma^j_{ih,k}\frac{\partial x^i}{\partial y^\ell}\frac{\partial x^h}{\partial y^m}\frac{\partial x^k}{\partial y^r} - \Gamma^j_{kh,i}\frac{\partial x^i}{\partial y^\ell}\frac{\partial x^h}{\partial y^m}\frac{\partial x^k}{\partial y^r} + \Gamma^j_{kf}\Gamma^f_{ih}\frac{\partial x^k}{\partial y^r}\frac{\partial x^i}{\partial y^\ell}\frac{\partial x^h}{\partial y^m} - \Gamma^j_{if}\Gamma^f_{kh}\frac{\partial x^i}{\partial y^\ell}\frac{\partial x^k}{\partial y^r}\frac{\partial x^h}{\partial y^m}\right).\end{aligned}$$

When we denote the curvature tensor in the y-coordinates by $S^q_{\ell rm}$ and recall (4.146), we can write this as

$$\frac{\partial^3 x^j}{\partial y^m \partial y^\ell \partial y^r} - \frac{\partial^3 x^j}{\partial y^m \partial y^r \partial y^\ell} = S^q_{\ell rm}\frac{\partial x^j}{\partial y^q} - R^j_{ikh}\frac{\partial x^i}{\partial y^\ell}\frac{\partial x^h}{\partial y^m}\frac{\partial x^k}{\partial y^r}. \tag{4.152}$$

Thus, the condition that the third derivatives commute becomes

$$S^q_{\ell rm} = R^j_{ikh}\frac{\partial x^i}{\partial y^\ell}\frac{\partial x^k}{\partial y^r}\frac{\partial x^h}{\partial y^m}\frac{\partial y^q}{\partial x^j}, \tag{4.153}$$

that is, the curvature tensor in the y-coordinates is the transform of the curvature tensor in the x-coordinates. This condition is equivalent to the condition (4.110) given above [see (4.18)], and it thus is a necessary condition for transforming the metric $h_{k\ell}dy^k dy^\ell$ into the metric $g_{ij}dx^i dx^j$. In fact, the conditions (4.153) are also sufficient for this in the following sense. As we have already explained above, according to the Theorem of Frobenius (1877), one can solve a system of equations of the form (4.124)

$$\frac{\partial g}{\partial x^j} = \phi_j \tag{4.154}$$

for given functions ϕ_j and an unknown function g in a simply connected domain if and only if the conditions for the commutativity of the second partial derivatives of g, $\frac{\partial^2 g}{\partial x^j \partial x^k} = \frac{\partial^2 g}{\partial x^k \partial x^j}$, that is,

$$\frac{\partial \phi_j}{\partial x^k} = \frac{\partial \phi_k}{\partial x^j} \text{ for all } j, k \tag{4.155}$$

are satisfied. And as before, (4.153) precisely provides the integrability conditions (4.155) for the differential Eq. (4.142) for the functions $\frac{\partial y^\ell}{\partial x^i}$. Thus, (4.153) is the necessary and sufficient local condition for the existence of the desired transformation, indeed.

Thus, we can conclude Riemann's fundamental result.

Theorem 4.2 *Two Riemannian metrics $g_{ij}dx^i dx^j$ and $h_{k\ell}dy^k dy^\ell$ can be transformed into each other by a change of coordinates if and only if their curvature tensors can be transformed into each other. Thus, a Riemannian metric is fully determined by its curvature tensor.*

Again, we point out that this is a local result, valid for transformations of local coordinates. No claim is made here about the global topology of a manifold. While the issue of the global topology also emerges from Riemann's work (more precisely, from his work on what now are called Riemann surfaces), the habilitation text does not address those issues.

Following Riemann, we now define

Definition 4.5 A Riemannian metric $g_{ij}dx^i dx^j$ is called *flat* if it can be transformed into the Euclidean metric $\delta_{k\ell}dy^k dy^\ell$.

We then have the following important special case of Theorem 4.2.

Corollary 4.1 *A metric is flat if and only if its curvature tensor vanishes.*

As mentioned in the beginning of this section, in his Paris essay, Riemann presented the calculation, which we have first described along Riemann's original lines and then worked out here in general terms for the proof of Theorem 4.2, only for flat metrics, that is, for the situation covered by Corollary 4.1. The corollary is the most interesting case of the theorem. While the general theorem only reduces the question of the equivalence of two metric tensors to that of two other tensors, their curvature tensors, the corollary makes use of the invariant properties of the latter, namely that for a flat metric, it is zero in arbitrary coordinates whereas the form of the metric tensor depends on those coordinates.

Putting it somewhat differently, from the curvature tensor, we can derive invariants, like the sectional curvatures to be introduced in Sect. 4.4.9, that do not change when we change the coordinates. The metric tensor itself, however, is not invariant under coordinate changes. It is this property that makes the curvature tensor so useful.

4.4.8 Submanifolds of Euclidean Space

Riemannian geometry removes a conceptual inconsistency of classical differential geometry that considered curves on surfaces in Euclidean space. In the context of Riemannian geometry, one can consider surfaces as two-dimensional Riemannian manifolds in general three- (or higher-)dimensional Riemannian manifolds, not just Euclidean spaces. That is, the submanifold now belongs to the same category as the manifold that it is contained in. We can thus systematically investigate what geometry the ambient manifold induces on a submanifold, and conversely, what the geometry of submanifolds tells us about the geometry of the ambient space. In particular, as we shall see, the curvatures of particular surfaces in a general Riemannian manifold determine the latter's geometry. This constitutes one of the fundamental insights of Riemann.

Thus, the geometry of surfaces in three-dimensional Euclidean space is a rather special case of Riemannian geometry. We shall now apply the general constructions of Riemannian geometry to this special case. This will lead us, in particular, to a conceptual explanation of the fundamental theorem of Gauss on the geometry of surfaces, what he called the *Theorema egregium*, stating that the Gauss curvature of a surface, even though defined in terms of its embedding into Euclidean space, does not depend on that embedding, but only on the intrinsic geometry of the surface in question.

We denote Euclidean three-space by $\mathbb{E}^3$. It is globally coordinatized by $\mathbb{R}^3 = \{(z^1, z^2, z^3)\}$, and it carries the Euclidean metric $\delta_{\alpha\beta}$,

$$\langle V, W\rangle = \delta_{\alpha\beta} v^\alpha w^\beta = v^1 w^1 + v^2 w^2 + v^3 w^3 \text{ for } V = v^\alpha \frac{\partial}{\partial z^\alpha}, W = w^\alpha \frac{\partial}{\partial z^\alpha}. \quad (4.156)$$

We then want to study the local geometry of immersed surfaces in $\mathbb{E}^3$. Thus, we consider an injective smooth mapping $f = (f^1, f^2, f^3)$ from some two-dimensional domain U with coordinates x^1, x^2 into $\mathbb{E}^3$. The vectors $X_1(x) := \frac{\partial f(x)}{\partial x^1}$, $X_2(x) := \frac{\partial f(x)}{\partial x^2}$ then are tangent vectors of the local surface $f(U)$ at the point $f(x)$ in $\mathbb{E}^3$. We put

$$g_{ij}(x) := \langle X_i(x), X_j(x)\rangle. \quad (4.157)$$

This is the restriction of the Euclidean metric of $\mathbb{E}^3$ to the surface $f(U)$. In other words, it is the metric on the surface induced by the ambient Euclidean metric. It thus defines a Riemannian metric g on U, with

$$g_{ij} =: g\left(\frac{\partial}{\partial x^i}, \frac{\partial}{\partial x^j}\right). \quad (4.158)$$

We point out that while the Euclidean tangent vectors $\frac{\partial}{\partial z^\alpha}$ are orthonormal, that is, $\langle \frac{\partial}{\partial z^\alpha}, \frac{\partial}{\partial z^\beta}\rangle = \delta_{\alpha\beta}$, this is not the case for the intrinsic tangent vectors $\frac{\partial}{\partial x^i}$ of the surface, and therefore g_{ij} in general is not the identity matrix.

Note that the definition of this metric involves only the tangent vectors of the surface $f(U)$, but not its normal vector which indicates the direction in $\mathbb{E}^3$ orthogonal to those tangent vectors.

The metric defined by (4.157) is evaluated on the surface $f(U)$ in $\mathbb{E}^3$, whereas (4.158) is defined on U. We can simply switch between these two perspectives by identifying $\frac{\partial}{\partial x^i}$ with $X_i(x) = \frac{\partial f(x)}{\partial x^i}$, that is, the intrinsic tangent vector on the two-dimensional domain U with its image in $\mathbb{E}^3$. The concrete metric on $f(U)$ induced by the Euclidean metric is identified with an abstract metric on the coordinate domain U.

The vector X_i is a vector in $\mathbb{E}^3$ and therefore, it has three components,

$$X_i = X_i^\alpha \frac{\partial}{\partial z^\alpha} = \frac{\partial f^\alpha(x)}{\partial x^i} \frac{\partial}{\partial z^\alpha}, \tag{4.159}$$

where α runs through the values 1, 2, 3. This also yields the classical formula for what is called the *first fundamental form* of the surface in $\mathbb{E}^3$,

$$g_{ij} = \sum_\alpha \frac{\partial f^\alpha(x)}{\partial x^i} \frac{\partial f^\alpha(x)}{\partial x^j}, \tag{4.160}$$

confirming again the remark that g_{ij} in general is not the identity matrix.

Moreover, in $\mathbb{E}^3$, we have the standard derivative

$$D_{\frac{\partial}{\partial z^\alpha}} \left(v^\beta(z) \frac{\partial}{\partial z^\beta} \right) = \frac{\partial v^\beta(z)}{\partial z^\alpha} \frac{\partial}{\partial z^\beta}, \tag{4.161}$$

which, of course, satisfies the rules of Definition 4.3. This then also induces a derivative on our surface (in the same way that the Euclidean metric on $\mathbb{E}^3$ induced a metric on U), via

$$D_{\frac{\partial}{\partial x^i}} \left(v^\beta((f(x)) \frac{\partial}{\partial z^\beta} \right) := D_{X_i} \left(v^\beta((f(x)) \frac{\partial}{\partial z^\beta} \right) = \frac{\partial v^\beta(f(x))}{\partial x^i} \frac{\partial}{\partial z^\beta}. \tag{4.162}$$

The problem, however, is that even if $V = v^\beta((f(x)) \frac{\partial}{\partial z^\beta}$ is a tangent vector to the surface $f(U)$, that is, a linear combination of X_1 and X_2, then $D_{\frac{\partial}{\partial x^i}} V$ need not be tangent to $f(U)$. It could have a nonvanishing component in the third direction, the direction normal to the surface. In particular, $D_{X_i} X_j$ need not be tangent. But there is a simple solution to this problem: Project $D_{\frac{\partial}{\partial x^i}} V(f(x))$ onto the tangent plane of $f(U)$ at $f(x)$. That tangent plane is spanned by X_1 and X_2, and so we put

$$\nabla_{\frac{\partial}{\partial x^i}} V = \nabla_{X_i} V := g^{k\ell} \langle D_{X_i} V, X_\ell \rangle X_k. \tag{4.163}$$

Theorem 4.3

$$\nabla_{X_i} X_j = \Gamma_{ij}^k X_k. \tag{4.164}$$

Thus, the projection of the Euclidean derivative onto the surface $f(U)$ coincides with the covariant derivative on that surface induced by the metric g_{ij}.

Proof We have

$$\frac{\partial}{\partial x^i} g_{j\ell} = \frac{\partial}{\partial x^i}\langle X_j, X_\ell\rangle = \langle D_{X_i} X_j, X_\ell\rangle + \langle X_j, D_{X_i} X_\ell\rangle$$

and hence

$$\frac{\partial}{\partial x^i} g_{j\ell} + \frac{\partial}{\partial x^j} g_{i\ell} - \frac{\partial}{\partial x^\ell} g_{ij} = 2\langle D_{X_i} X_j, X_\ell\rangle$$

and so

$$\nabla_{X_i} X_j = g^{k\ell}\langle D_{X_i} X_j, X_\ell\rangle X_k = \Gamma_{ij}^k X_k \text{ from (4.61).}$$

□

I should point out that the presentation given here reverses the historical order of the discoveries. Riemann's followers like Christoffel and Levi-Civita first found that one can define an intrinsic notion of derivative by (4.164) for submanifolds of Euclidean spaces. A version of Definition 4.3 was only formulated later.

We now return to the Euclidean derivative (4.161) and compare it with its projection (4.163) onto the surface. That projection simply discards the direction normal to the surface. Thus, if we let ν be a unit normal vector to the surface at the point under consideration, that is

$$\langle \nu(f(x)), X_i(f(x))\rangle = 0 \text{ for } i = 1, 2, \text{ and } \langle \nu(f(x)), \nu(f(x))\rangle = 1, \tag{4.165}$$

we may write

$$D_{X_i} X_j = \nabla_{X_i} X_j + h_{ij}\nu, \tag{4.166}$$

where the coefficients h_{ij} of the *second fundamental form* are given by

$$\langle D_{X_i} X_j, \nu\rangle = -\langle X_j, D_{X_i}\nu\rangle \tag{4.167}$$

by differentiating the identity $\langle X_j, \nu\rangle = 0$ from (4.165). Moreover, since X_ℓ is tangential to the surface, we have $\langle D_{X_i}\nabla_{X_j} X_k, X_\ell\rangle = \langle \nabla_{X_i}\nabla_{X_j} X_k, X_\ell\rangle$, and therefore differentiating (4.167) yields

$$\langle D_{X_i} D_{X_j} X_k, X_\ell\rangle = \langle D_{X_i}\nabla_{X_j} X_k + D_{X_i}(h_{jk}\nu), X_\ell\rangle = \langle \nabla_{X_i}\nabla_{X_j} X_k, X_\ell\rangle - h_{jk}h_{i\ell}. \tag{4.168}$$

Since $D_{X_i} D_{X_j} = D_{X_j} D_{X_i}$ (covariant derivatives commute in the flat Euclidean space), we obtain

Theorem 4.4 (Theorema egregium of Gauss) *The Gauss curvature of the surface $f(U)$ in $\mathbb{E}^3$ satisfies*

$$\frac{R_{1212}}{g_{11}g_{22} - g_{12}^2} = \frac{\langle (D_{X_1}D_{X_2} - D_{X_2}D_{X_1})X_2, X_1\rangle}{g_{11}g_{22} - g_{12}^2} = \frac{h_{11}h_{22} - h_{12}^2}{g_{11}g_{22} - g_{12}^2} =: K \tag{4.169}$$

Of course, the first identity is simply the definition of the sectional curvature which we shall present in a moment in Sect. 4.4.9, and the last identity is the classical definition of the Gauss curvature K in terms of the second fundamental form. Thus, the definition seems to depend not only on the intrinsic geometry of the surface as encoded by the first fundamental form, that is, its Riemannian metric, but also on the second fundamental form, that is, how the surface sits in $\mathbb{E}^3$. Gauss' fundamental discovery then was that K can be expressed in solely intrinsic terms, that is, from derivatives of the g_{ij}. Riemann recovers this result and puts it in the general context of his concept of curvature. In fact, analogous constructions apply to submanifolds of Euclidean spaces of any dimensions. Moreover, the Gauss curvature can be defined in this manner, as the left hand side of (4.169), for any surface in any Riemannian manifold. This will provide Riemann with an important interpretation of the sectional curvature introduced in Sect. 4.4.9.

4.4.9 Sectional Curvature

We now wish to explain another fundamental insight of Riemann. In Sect. 4.4.6, we have presented Riemann's argument that $\frac{n(n-1)}{2}$ scalar quantities should suffice to uniquely determine a Riemannian metric up to coordinate transformations. In Theorem 4.2, we have seen that the curvature tensor determines the metric up to coordinate transformations. When looking at (4.146) or (4.147), the curvature tensor seems to possess more than $\frac{n(n-1)}{2}$ independent components, however. As we shall discuss in this section, this is not so. The algebraic symmetries that the curvature tensor possesses reduce the number of independent components to $\frac{n(n-1)}{2}$. Moreover, these independent components can be identified with the curvatures in the sense of Theorem 4.4 of certain surfaces in the manifold. Since there are precisely $\frac{n(n-1)}{2}$ independent surface directions in an n-dimensional manifolds, we obtain the predicted number of parameters, and moreover, this then fits with the interpretation of curvature provided in Theorem 4.1.

We now derive the symmetries of the curvature tensor. These are precisely the symmetries that we have already found above in (4.101)–(4.104) of Sect. 4.4.5.

Lemma 4.2 *For vector fields X, Y, Z, W, we have*

$$R(X,Y)Z = -R(Y,X)Z, \quad \text{i.e. } R_{k\ell ij} = -R_{k\ell ji}, \tag{4.170}$$

$$R(X,Y)Z + R(Y,Z)X + R(Z,X)Y = 0, \quad \text{i.e. } R_{k\ell ij} + R_{kij\ell} + R_{kj\ell i} = 0, \tag{4.171}$$

$$\langle R(X,Y)Z, W\rangle = -\langle R(X,Y)W, Z\rangle, \quad \text{i.e. } R_{k\ell ij} = -R_{\ell kij}, \tag{4.172}$$

$$\langle R(X,Y)Z, W\rangle = \langle R(Z,W)X, Y\rangle, \quad \text{i.e. } R_{k\ell ij} = R_{ijk\ell}. \tag{4.173}$$

Equation (4.171) is called the first Bianchi identity.

Proof Since we are dealing with a tensor, it suffices to verify all claims for coordinate vector fields $\frac{\partial}{\partial x^i}$. Since coordinate vector fields commute, their Lie brackets vanish. We may thus assume that all Lie brackets (4.138) of X, Y, Z and W vanish. Equation (4.170) then follows directly from (4.146). For (4.171), we observe

$$\begin{aligned} R(X, Y)Z + R(Y, Z)X + R(Z, X)Y \\ &= \nabla_X \nabla_Y Z - \nabla_Y \nabla_X Z + \nabla_Y \nabla_Z X \\ &\quad - \nabla_Z \nabla_Y X + \nabla_Z \nabla_X Y - \nabla_X \nabla_Z Y \\ &= 0, \end{aligned}$$

since $\nabla_Y Z = \nabla_Z Y$ etc. by (4.136).

For (4.172), we shall show $\langle R(X, Y)Z, Z\rangle = 0$ for all X, Y, Z, i.e. $R_{kkij} = 0$.

$$\begin{aligned} \langle \nabla_X \nabla_Y Z, Z\rangle &= X\langle \nabla_Y Z, Z\rangle - \langle \nabla_Y Z, \nabla_X Z\rangle \text{ since } \nabla \text{ is metric} \\ &= \frac{1}{2} XY\langle Z, Z\rangle - \langle \nabla_Y Z, \nabla_X Z\rangle \text{ for the same reason} \\ &= \frac{1}{2} YX\langle Z, Z\rangle - \langle \nabla_Y Z, \nabla_X Z\rangle \text{ assuming again w.l.o.g. } [X, Y] = 0 \\ &= \langle \nabla_Y \nabla_X Z, Z\rangle + \langle \nabla_Y Z, \nabla_X Z\rangle - \langle \nabla_Y Z, \nabla_X Z\rangle \text{ going backwards} \\ &= \langle \nabla_Y \nabla_X Z, Z\rangle, \end{aligned}$$

which, recalling that we assume $[X, Y] = 0$, implies that $\langle R(Y, X)Z, Z\rangle$ vanishes, as desired.

It is left to prove (4.173), which as we shall see, is a consequence of the other symmetries. From (4.170), (4.171)

$$\begin{aligned} \langle R(X, Y)Z, W\rangle &= -\langle R(Y, X)Z, W\rangle \\ &= \langle R(X, Z)Y, W\rangle + \langle R(Z, Y)X, W\rangle, \end{aligned} \tag{4.174}$$

and from (4.171), (4.172)

$$\begin{aligned} \langle R(X, Y)Z, W\rangle &= -\langle R(X, Y)W, Z\rangle \\ &= \langle R(Y, W)X, Z\rangle + \langle R(W, X)Y, Z\rangle. \end{aligned} \tag{4.175}$$

From (4.174), (4.175)

$$\begin{aligned} 2\langle R(X, Y)Z, W\rangle = \langle R(X, Z)Y, W\rangle + \langle R(Z, Y)X, W\rangle \\ + \langle R(Y, W)X, Z\rangle + \langle R(W, X)Y, Z\rangle. \end{aligned} \tag{4.176}$$

Analogously,

$$\begin{aligned}2\langle R(Z,W)X,Y\rangle &= \langle R(Z,X)W,Y\rangle + \langle R(X,W)Z,Y\rangle \\ &\quad + \langle R(W,Y)Z,X\rangle + \langle R(Y,Z)W,X\rangle \\ &= 2\langle R(X,Y)Z,W\rangle\end{aligned}$$

by applying (4.170) and (4.172) to all terms. □

Alternatively, we could have derived these symmetries directly from the formula for the curvature in normal coordinates, that is, (4.149) or (4.100). Since the symmetries that we have obtained in Lemma 4.2 are the same that we had already found in Sect. 4.4.5, this suggests that we return to the interpretation of the curvature tensor given there. That interpretation involved surface elements, and we now formulate the abstract concept.

Definition 4.6 The *sectional curvature* of the plane spanned by the (linearly independent) tangent vectors $V = v^i \frac{\partial}{\partial x^i}$, $W = w^i \frac{\partial}{\partial x^i} \in T_x M$ of the Riemannian manifold M is

$$\begin{aligned}K(V\wedge W) :&= \langle R(V,W)W,V\rangle \frac{1}{|V\wedge W|^2} \\ &= \frac{R_{ijk\ell} v^i w^j v^k w^\ell}{g_{ik}g_{j\ell}(v^i v^k w^j w^\ell - v^i v^j w^k w^\ell)} \\ &= \frac{R_{ijk\ell} v^i w^j v^k w^\ell}{(g_{ik}g_{j\ell} - g_{ij}g_{k\ell}) v^i w^j v^k w^\ell}\end{aligned} \tag{4.177}$$

($|V\wedge W|^2 = \langle V,V\rangle\langle W,W\rangle - \langle V,W\rangle^2$).

Remark 4.1

1. The sectional curvature of the plane $V \wedge W$ is a scalar quantity, being the quotient of two tensors, the curvature tensor and the square of the metric tensor. The numerator and the denominator are both evaluated twice on the vectors V and W, and therefore the transformation factors in the numerator and the denominator cancel. Thus, the sectional curvatures are invariant under coordinate changes. Only scalar quantities can yield invariants, as (other) tensors transform nontrivially under coordinate changes. Thus, importantly, we do not get invariants from either the metric or the curvature tensor alone, but we need to form a suitable quotient in order to obtain scalar invariants.
2. There is an important exception, however: When the curvature tensor vanishes, then it is simply zero regardless of what the metric tensor looks like in the chosen coordinates. This is the case of a flat metric, and so, we can detect flatness in arbitrary coordinates by simply looking at the curvature tensor.

3. The curvature tensor essentially captures the geometric information contained in the second derivatives of the metric tensor, see Sect. 4.4.5. Thus, the sectional curvatures are essentially a quotient of the second derivatives of the metric by a quadratic expression in the metric tensor itself.
4. The denominator in (4.177) is the area of the parallelogram spanned by V and W in the tangent space under consideration. Equivalently, it is twice the area of the triangle with vertices $0, V, W$. As already seen in Sect. 4.4.5, the sectional curvature expresses the deviation of the area of an infinitesimal triangle in the manifold from that in the tangent space.

We can now easily verify the fundamental result of Riemann that the sectional curvatures determine the whole curvature tensor, or more precisely, all the quantities $\langle R(V, W)Z, Y\rangle$ (note that these quantities involve the metric tensor).

Lemma 4.3 *With*

$$K(V, W) := K(V \wedge W)|V \wedge W|^2 = \langle R(V, W)W, V\rangle, \tag{4.178}$$

we have

$$\begin{aligned}\langle R(V, W)Z, Y\rangle = \frac{1}{6}\Big(&K(V+Y, W+Z) - K(V+Y, W) - K(V+Y, Z)\\ &- K(V, W+Z) - K(Y, W+Z) + K(V, Z) + K(Y, W)\\ &- K(W+Y, V+Z) + K(W+Y, V) + K(W+Y, Z)\\ &+ K(W, V+Z) + K(Y, V+Z) - K(W, Z) - K(Y, V)\Big).\end{aligned} \tag{4.179}$$

Proof This follows by a direct computation from Lemma 4.2, systematically inserting (4.178) into (4.179). □

From (4.149), we also see that in normal coordinates, we have the simple expression

$$K\left(\frac{\partial}{\partial x^i}, \frac{\partial}{\partial x^j}\right) = \left\langle R\left(\frac{\partial}{\partial x^i}, \frac{\partial}{\partial x^j}\right)\frac{\partial}{\partial x^j}, \frac{\partial}{\partial x^i}\right\rangle = R_{jiji} = \frac{1}{2}\left(2g_{ij,ij} - g_{ii,jj} - g_{jj,ii}\right). \tag{4.180}$$

A version of this formula occurs in Riemann's Paris essay.

Riemann provides the following geometric interpretation of the sectional curvatures. At the point p of the Riemannian manifold M under consideration, take a tangent plane spanned by two independent vectors V and W and consider the surface in M formed by all geodesics starting at p with initial directions in that tangent plane, that is, with initial directions that are linear combinations of V and W. The Gauss curvature of this surface at p then is the sectional curvature of the tangent plane spanned by V and W.

We can therefore utilize the properties of the Gauss curvature to obtain the geometric interpretation of sectional curvature. We shall proceed to do so. We start with the metric in geodesic polar coordinates (4.83), which we write as

$$dr^2 + m^2 d\varphi^2. \tag{4.181}$$

Since the metric in normal coordinates is diagonal at the origin, and all first derivatives vanish, we obtain from the formula for transforming those coordinates into polar coordinates that

$$m(0) = 0, \quad \frac{\partial m}{\partial r}(0) = 1. \tag{4.182}$$

When we let the index 1 correspond to r and 2 to φ, we have

$$g_{11} = 1, \quad g_{12} = 0, \quad g_{22} = m^2. \tag{4.183}$$

By Definition 4.2 and (4.177), the sectional curvature of our plane then is

$$K := \frac{R_{1212}}{g_{11}g_{22} - g_{12}^2} = \frac{1}{m^2}\left(\Gamma^1_{22,1} - \Gamma^1_{12,2} + \Gamma^1_{p1}\Gamma^p_{22} - \Gamma^1_{p2}\Gamma^p_{12}\right). \tag{4.184}$$

From (4.183), and denoting a derivative of m by a subscript, for instance $m_1 = \frac{\partial m}{\partial r}$, we also compute (4.62),

$$\Gamma_{122} = -mm_1, \quad \Gamma_{212} = \Gamma_{221} = mm_1, \quad \Gamma_{222} = mm_2, \text{ all other } \Gamma^k_{ij} = 0.$$

Thence

$$\Gamma^1_{22,1} = \Gamma_{122,1} = -(m_1^2 + mm_{11}), \quad \Gamma^1_{22}\Gamma^2_{12} = \Gamma_{122}\frac{1}{m^2}\Gamma_{212} = -m_1^2,$$

while all other terms in (4.184) vanish. Thus, the sectional curvature is simply

$$K = -\frac{m_{11}}{m}. \tag{4.185}$$

Of course, this becomes undefined at $r = 0$, but we can obviously take the limit $r \to 0$, as when the curvature is continuous there, we also need to have $m_{11}(0) = 0$, in addition to the relations $m(0) = 0$, $m_1(0) = 1$ from (4.182), and hence by L'Hôpital's rule

$$K(0) = -\frac{\partial^3 m}{\partial r^3}(0). \tag{4.186}$$

This yields the expansion

$$m = r - K(0)\frac{r^3}{6} + o(r^3). \tag{4.187}$$

Therefore, the length $L(\rho)$ of the circle $r \equiv \rho$ can be expanded as

$$L(\rho) = \int_0^{2\pi} m d\varphi = 2\pi\rho - K(0)\frac{\pi\rho^3}{3} + o(\rho^3), \tag{4.188}$$

yielding the formula

$$K(0) = \frac{3}{\pi} \lim_{\rho\to 0} \frac{2\pi\rho - L(\rho)}{\rho^3} \tag{4.189}$$

as a geometric interpretation of curvature in terms of the length of the circle at distance ρ from the point under consideration. Similarly, when we denote by $A(\rho)$ the area of the disc $\{r \leq \rho\}$, we have

$$K(0) = \frac{12}{\pi} \lim_{\rho\to 0} \frac{\pi\rho^2 - A(\rho)}{\rho^4}. \tag{4.190}$$

We now derive Gauss' formula, called *Theorema elegantissimum* by him, for the relation between curvature and angle sum in a geodesic triangle which also comes up in Riemann's text.

Theorem 4.5 (Theorema elegantissimum of Gauss) *Let Δ be a triangle contained in some coordinate neighborhood whose three sides are geodescic arcs, with angles $\theta_1, \theta_2, \theta_3$ at the vertices. Let K denote the Gauss curvature. Then*

$$\int_\Delta K m dr d\varphi = \sum_{j=1}^{3} \theta_j - \pi. \tag{4.191}$$

Of course, this includes the Euclidean result that the sum of the angles in a triangle is π. $mdrd\varphi$ is simply the area element, and so Gauss' Theorem says that the deviation of the angle sum from π in a geodesic triangle is measured by the integral of the Gauss curvature in that triangle.

Remark 4.2 This result also holds globally, without the restriction that the triangle be contained in a geodesic polar coordinate region, and also for other geodesic polygons. This result is called the *Theorem of Gauss-Bonnet*. The general result can be readily obtained from the preceding one by decomposing a geodesic polygon into sufficiently small geodesic triangles.

We sketch the *Proof.* We assume (for purely technical reasons, see the preceding remark) that the triangle is so small that we can introduce geodesic polar coordinates at each vertex so that Δ is contained in that coordinate region. From (4.185), we get in geodesic polar coordinates

$$\int_{\Delta} K m dr d\varphi = -\int_{\Delta} \frac{\partial^2 m}{\partial r^2} dr d\varphi$$
$$= -\int_{\partial\Delta} \frac{\partial m}{\partial r} d\varphi.$$

We take geodesic polar coordinates at a vertex P_j. Thus, two of the sides of Δ then are radial geodesics emanating from P_j. On each of these two sides, φ thus is constant, and so, $d\varphi$ vanishes there. However, in order to evaluate the above integral, we also need to consider the jump that φ makes at P_j when passing from one to the other side. This jump is the outer angle at P_j, that is, $\pi - \theta_j$. Since $\frac{\partial m}{\partial r}(0) = 1$ by (4.182), this then also is the contribution of the jump in the integral. Thus, in the above integral, we do not get any contribution from the sides, but just contributions from the vertices. But since we turn around by a total angle of 2π when we traverse the triangle once, we need to subtract 2π in our accounting. This means that

$$\int_{\partial\Delta} \frac{\partial m}{\partial r} d\varphi = \sum_j (\pi - \theta_j) - 2\pi$$

which yields the result. □

We return to the consideration of general Riemannian manifolds. The curvature tensor satisfies one further symmetry, the second Bianchi identity. Although this identity does not yet appear in Riemann's work, we present it here as it reflects an important property of the Riemann curvature tensor.

Lemma 4.4 (Second Bianchi Identity)

$$\nabla_{\frac{\partial}{\partial x^h}} R_{k\ell ij} + \nabla_{\frac{\partial}{\partial x^k}} R_{\ell h ij} + \nabla_{\frac{\partial}{\partial x^\ell}} R_{hkij} = 0. \tag{4.192}$$

Proof We work in normal coordinates at the point under consideration and utilize the formula (4.149) for the components of the curvature tensor in normal coordinates,

$$R_{k\ell ij} = \frac{1}{2}(g_{jk,\ell i} + g_{i\ell,kj} - g_{j\ell,ki} - g_{ik,\ell j}).$$

Since we all terms involving first derivatives of the metric vanish in normal coordinates, we then also get

$$R_{k\ell ij,h} = \frac{1}{2}(g_{jk,\ell ih} + g_{i\ell,kjh} - g_{j\ell,kih} - g_{ik,\ell jh}). \tag{4.193}$$

Thus

$$\begin{aligned}R_{k\ell ij,h} + R_{\ell hij,k} + R_{hkij,\ell} &= \frac{1}{2}\big(g_{jk,\ell ih} + g_{i\ell,kjh} - g_{j\ell,kih} - g_{ik,\ell jh}\\ &\quad + g_{j\ell,hik} + g_{ih,\ell jk} - g_{jh,\ell ik} - g_{i\ell,hjk}\\ &\quad + g_{jh,ki\ell} + g_{ik,hj\ell} - g_{jk,hi\ell} - g_{ih,kj\ell}\big)\\ &= 0.\end{aligned}$$

□

4.4.10 Spaces of Constant Curvature

Definition 4.7 The Riemannian manifold M is called a space of constant sectional curvature, or a *space form*, if $K(X \wedge Y) = K \equiv$ const. for all linearly independent $X, Y \in T_x M$ and all $x \in M$. A space form is called *spherical*, *flat*, or *hyperbolic*, depending on whether $K > 0, = 0, < 0$.

From Corollary 4.1 and Lemma 4.3, we know that the Riemannian manifolds of vanishing sectional curvature, the *flat* ones, are those that possess local coordinates for which the coordinate vector fields $\frac{\partial}{\partial x^i}$ satisfy

$$g_{ij} = \left\langle \frac{\partial}{\partial x^i}, \frac{\partial}{\partial x^j} \right\rangle \equiv \delta_{ij}.$$

From the second Bianchi identity, we shall now deduce the Theorem of Issai Schur (1875–1941).

Theorem 4.6 (Schur) *Let the dimension n of the Riemannian manifold be at least 3. If the sectional curvature of M is constant at each point, i.e.*

$$K(X \wedge Y) = f(x) \quad \text{for } X, Y \in T_x M, \tag{4.194}$$

then $f(x) \equiv$ const and M is a space form.

Proof Let K be constant at every point x, i.e. $K(X \wedge Y) = f(x)$. From Lemma 4.3, we obtain

$$R_{ijk\ell} = f(x)(g_{i\ell} g_{jk} - g_{ik} g_{j\ell}).$$

By Lemma 4.4, with normal coordinates at x, and with $f_h := \frac{\partial}{\partial x^h}(f)$, we obtain

$$
\begin{aligned}
0 &= R_{ijk\ell,h} + R_{jhk\ell,i} + R_{hik\ell,j} \\
&= f_h(\delta_{i\ell}\delta_{jk} - \delta_{ik}\delta_{j\ell}) + f_i(\delta_{j\ell}\delta_{hk} - \delta_{jk}\delta_{h\ell}) + f_j(\delta_{h\ell}\delta_{ik} - \delta_{hk}\delta_{i\ell}).
\end{aligned}
$$

Since we assume $\dim M \geq 3$, for each h, we can find i, j, k, ℓ with $i = \ell, j = k, h \neq i, h \neq j, i \neq j$. It follows that $0 = f_h$. Since this holds for all $x \in M$ and all h, we conclude $f \equiv$ const. □

Schur's theorem is remarkable, because a pointwise property implies a global one. If at each point all directions are geometrically indistinguishable, then also all points are geometrically indistinguishable. Expressed in a general terminology, an isotropic Riemannian manifold is homogeneous.

After having discussed Schur's theorem, which should be helpful for the general perspective, we return to Riemann's reasoning.

From the fact that the sectional curvatures (in $\frac{n(n-1)}{2}$ general surface directions) determine the metric, that is, the local geometry, we can now immediately draw important consequences. When the curvature is the same in every surface direction, then also the geometry is the same in any such direction. And if the curvature is the same at different points (which, incidentally, by Schur's theorem already follows when the curvature at each point is the same in every surface direction), then also the geometry is the same at those points. Therefore, in a space of constant curvature, any body can be freely moved and rotated without distortions. (In modern terminology, for any two points in such a manifold, there exist local isometries between suitable neighborhoods of them.) Conversely, if we can freely move a body without distortion, then the geometry must be the same at each point. In particular, the areas of geodesic triangles need to coincide, as otherwise, there would be distortions of local surfaces. But we have seen in Theorem 4.1 in Sect. 4.4.5 that such areas of triangles are given in terms of the curvature. Therefore, for bodies to move freely, we need to have constant curvature.

This reasoning is very simple and beautiful. It remains to explicitly give a Riemannian metric of constant curvature k, for any $k \in \mathbb{R}$. Riemann states a formula for such a metric in local coordinates, and we shall now explain and derive that formula.

We write the metric again in geodesic polar coordinates (4.83)

$$dr^2 + m^2 d\varphi^2, \tag{4.195}$$

and where we treat φ as one-dimensional, because for the sectional curvature, it suffices to consider surfaces. By (4.185), the sectional curvature of our plane then is

$$k = -\frac{m_{11}}{m}. \tag{4.196}$$

Of course, this becomes undefined at $r = 0$, but we can obviously take the limit $r \to 0$, as when the curvature is continuous there, we also need to have $m(0) = 0, m_1(0) = 1, m_{11}(0) = 0$ [see (4.182)]. When k is constant, the solution of (4.196) then is

$$m_k(r) = \begin{cases} \frac{\sin(\sqrt{k}r)}{\sqrt{k}} & \text{for } k > 0 \\ r & \text{for } k = 0 \\ \frac{\sinh(\sqrt{-k}r)}{\sqrt{-k}} & \text{for } k < 0. \end{cases} \tag{4.197}$$

Thus, in geodesic polar coordinates, a constant curvature metric is of the form

$$dr^2 + m_k(r)^2 d\varphi^2 \tag{4.198}$$

with m_k from (4.197).

We can now apply this to constant curvature metrics in dimension n and derive Riemann's formula. Since constant curvature metrics look the same in every surface direction, we can then let φ stand for the collection $\xi^1, \ldots, \xi^n$ with $\sum(\xi^i)^2 = 1$ of directional variables on the sphere S^{n-1} in (4.198), as in Sect. 4.4.5. For $k > 0$, we then put

$$x^i := \frac{2\xi^i}{\sqrt{k}} \tan \frac{\sqrt{k}r}{2} \tag{4.199}$$

to get

$$\begin{aligned} \sum(x^i)^2 &= \frac{4}{k}\tan^2\frac{\sqrt{k}r}{2} \\ \sum(dx^i)^2 &= \frac{dr^2}{\cos^4\frac{\sqrt{k}r}{2}} + \frac{4}{k}\tan^2\frac{\sqrt{k}r}{2}\sum(d\xi^i)^2 \\ ds &= \cos^2\frac{\sqrt{k}r}{2}\sqrt{\sum(dx^i)^2} \\ &= \frac{1}{1+\frac{k}{4}\sqrt{\sum(x^i)^2}}\sqrt{\sum(dx^i)^2}. \end{aligned} \tag{4.200}$$

We note that (4.199) is the stereographic projection of the sphere of radius $\sqrt{k}$ as in (4.10). Here, we take the north pole as the center of our coordinates. We observe that r then is the distance from the north pole measured on the sphere whereas the φ^i parametrize an $(n-1)$-dimensional subsphere centered at the north pole. In particular, for $k = 1$, the metric tensor is the same as that given in (4.43). Thus, the metric on the sphere S^n induced by that of the ambient Euclidean space of dimension $n+1$ has sectional curvature $\equiv 1$. It is geometrically clear that in such a sphere, we can move and rotate bodies without distortion, a crucial property of constant curvature spaces pointed out by Riemann. The case of other positive k thus is simply obtained by a scaling of the unit sphere.

In the case of $k < 0$, we use

$$x^i := \frac{2\xi^i}{\sqrt{-k}} \tanh \frac{\sqrt{-k}r}{2} \tag{4.201}$$

to arrive at the same formula (4.200). Of course, this formula is also valid for $k = 0$. (4.200) is the formula given by Riemann for a constant curvature metric. In the case of $k < 0$, we need to restrict the values of x to the open ball $\{\sum(x^i)^2 < \frac{4}{(-k)}\}$. In particular, we can realize a Riemannian metric of curvature -1 on the open unit ball.

4.5 Going Through Riemann's Text

Equipped with the preceding, we can now go through Riemann's text and understand his arguments.

In the introduction, Riemann explains that he wants to clarify the relation between axioms and geometric constructions. He says that what is needed is the concept of a multiply extended magnitude. On such an object, different notions of measurement can be introduced. Which of them is realized in our ordinary physical space, which is a threefold extended magnitude, then cannot be deduced a priori, but needs to be found empirically. The most important, but definitely not the only example is Euclidean space.

In Chap. 1, he then develops that notion of an n-fold extended magnitude. (Clifford translates Riemann's German word "Mannigfaltigkeit" as *manifoldness*, which is the most literal translation, but in the sequel, I shall use the simpler *manifold* which is the term that is by now generally established.) Riemann says that he can only draw upon the work of Gauss on the differential geometry of surfaces and on some work by the philosopher Herbart, but apart from that, he has to explore a completely new territory. Such manifolds arise naturally when an object can admit different qualifications, like the different positions of a point in space or the different colors of an object. Such manifolds could be either discrete, whence quanta can be counted, or continuous, the realm of measurement, and in his text, Riemann will investigate the latter case, as discussed in Sect. 4.4.1. Such continuous manifolds are also useful in mathematics, for the treatment of multi-valued analytic functions or for differential equations. n-dimensional manifolds are locally represented by n independent coordinate functions. In §2, he iteratively constructs a manifold of dimension $n+1$ from one of dimension n, by adding a degree of freedom. In §3, conversely, he explains that generically the level set of a function on an n-dimensional manifold is an $(n-1)$-dimensional submanifold. From these considerations, the dimension of a manifold can be built up iteratively. (As I have mentioned on p. 54, there are some technical difficulties associated with the concept of the dimension of a manifold. These difficulties disappear when we assume the manifold to be differentiable, as Riemann implicitly does in the sequel and as we have systematically done in our commentary.) There also exist infinite dimensional manifolds, like spaces of functions in a given region or the possible shapes of a solid figure.

In Chap. 2, he introduces metrics as described in Sect. 4.4.3 as a tool for performing measurements on a manifold. A Riemannian metric enables one to measure the length of a curve independently of its positions and thereby to compare the lengths of different curves. In his habilitation address, Riemann will not present the formulae needed for a rigorous development of the theory, but only explain the geometric results in abstract terms. In §1, the concept of what is now called a Riemannian metric is introduced, as treated in Sect. 4.4.3. While more general notions are possible, Riemann justifies his choice as the formally simplest, but contentful and nontrivial generalization of the Euclidean metric. The Euclidean metric is called *flat*, and much of his subsequent reasoning is concerned with the criterion for when a metric given in arbitrary coordinates is flat. As the metric on an n-dimensional manifold is given by a symmetric tensor $(g_{ij})_{i,j=1,\dots,n}$, there are $\frac{n(n+1)}{2}$ degrees of freedom. n of those are only apparent, as we can choose the n coordinate functions. Thus, there remain $\frac{n(n-1)}{2}$ real degrees of freedom. In §2, he then introduces normal coordinates as described in Sect. 4.4.5 and formulates the result of Theorem 4.1, and he identifies the crucial term in the expansion with the Gauss curvature. (Riemann employs a somewhat different normalization, and so he gets a factor $-\frac{3}{4}$.) Thus, the expansion of the metric to second order is determined by the Gauss curvatures of surfaces. Since at a point in an n-dimensional manifold, there are $\frac{n(n-1)}{2}$ independent surface directions, this gives precisely the right number of degrees of freedom, according to his count. That is, in general, a Riemannian metric is determined when we know $\frac{n(n-1)}{2}$ independent surface curvatures at each point. As explained in Sect. 4.4.6, one needs to read his text carefully here, as these surface directions need to be general and independent (in §4, he also makes the somewhat cryptic remark "whose curvature measures are independent of each other", which is not reproduced in any of the three English translations, including Clifford's).

In §3, he recalls the geometric interpretations of Gauss curvature, as given by Gauss in his Theorema egregium 4.4 as the product of the two principal curvatures of a surface in space and in his Theorema elegantissimum 4.5 in terms of the deviation of the sum of the angles in a geodesic triangle from π (see Sect. 4.4.4 for the notion of a geodesic curve; a geodesic triangle is one whose three sides are all geodesic). Thus, when he constructs a surface in a manifold from all the geodesic emanating from a given point in directions given by linear combinations of two tangent vectors, that is, in all the directions given by a two-dimensional plane in the tangent space, the curvature of that surface can be interpreted geometrically.

Thus, the sectional curvatures determine the metric. In particular, in §4, he says that flat metrics are characterized by the vanishing of their sectional curvatures, a theorem that we have derived in Sect. 4.4.7. More generally, (as described in Sect. 4.4.10) metrics of constant sectional curvature are precisely those in which bodies can be arbitrarily moved around without distortion, and he gives the formula (4.200) for such a constant curvature metric.

In §5, he describes the local geometry of surfaces of constant curvature. The reference surface is the unit sphere in Euclidean 3-space, which has constant curvature 1 (see Sect. 4.4.10). We then take pieces of surfaces of other constant curvature values k that touch that sphere along the equator. Such surfaces with $k > 1$ touch the equator from the inside, like the outer curve of a suitable torus of revolution. For $0 < k < 1$, take a larger sphere (of radius $\frac{1}{\sqrt{k}}$, cut out the piece between two suitable great half circles through the poles, with the distance of their intersection points with the equator being 2π, and glue those two half circles together to form a surface that can be made to touch the equator of the unit sphere from the outside. This surface then lies between the unit sphere and a vertical cylinder touching it also at the equator; the latter has curvature 0, as follows for instance from the Theorema egregium 4.4. Finally, a piece of surface of negative curvature touching the unit sphere at the equator would lie outside that cylinder, looking like the region around the inner curve of a suitable torus of revolution. In a surface of constant curvature, two-dimensional figures can be moved around with stretching. Since surfaces of positive curvature can be realized as spheres in Euclidean 3-space, figures can then also moved around in them without bending. Moreover, Riemann says that zero curvature surfaces are distinguished by the fact that in them also directions are independent of position. He seems to hint at the fact that in Euclidean space, we can unambigously identify the tangent spaces at different points by parallel transport. While in other spaces, one can also develop a notion of parallel transport (based on the covariant derivative introduced in Sect. 4.4.7), this parallel transport from one point to another will then depend on the choice of a connecting curve joining these two points, and therefore not be unambiguous or canonical.

Chapter 3 is devoted to applications to (physical) space. In §1, he provides various criteria for a Riemannian metric to be flat, that is, Euclidean. First, as developed in Sect. 4.4.7, we have the vanishing of the sectional curvatures in three independent and general surface directions at each point. According to the Theorema elegantissimum 4.5, this holds if the sum of the angles in any geodesic triangle is π. Secondly, when bodies are freely movable, the curvature is constant, and in order to find its value, we only need to check the sum of the angles in a single geodesic triangle, again appealing to the Theorema elegantissimum 4.5. Finally, as explained in my comment on §5 of the preceding chapter, one could also require that not only the lengths, but also the directions of curves do not depend on their position.

In §2, he argues that measurements in continuous manifolds can never be exact. Therefore, the determination of a Riemannian metric, say through its curvature, by measurements remains always inaccurate. This is an issue when going to smaller scales, for instance with the help of a microscope. When going to larger scales, up to the infinitely large, the distinction between *infinite extent* and *unboundedness*, as explained at the end of Sect. 4.4.1, becomes relevant. While we can reasonably assume that physical space is unbounded, we do not know whether it is also of infinite extent. In fact, should physical space have constant positive curvature, its diameter would necessarily be finite.

in §3, Riemann argues that while questions about the infinitely great are useless for the interpretation of nature, this is by no means so for the infinitely small. Progress in mechanics depends on the infinitesimal calculus and on the principles discovered by Archimedes, Galileo, and Newton, and for other, less exact sciences, the microscope offers us a glimpse into small scales for detecting the basic principles. When the metric of physical space has constant curvature, astronomical measurements so far do not indicate a deviation from the value 0. But when we abandon the assumption that bodies can freely move, that is, according to the preceding, that the curvature be constant, there can be deviations of the curvature from 0 in the small. These cannot be detected by our measurements which, because of their finite precision, can only yield local averages.

Riemann then makes the visionary remark that we cannot be certain that the empirical notions on which our metrical determinations of space are based, that of a solid body and of a light ray, are still valid in the infinitely small. Thus, the infinitesimal structure of space could in principle be very different. The question of the validity of the hypotheses of geometry is linked to that of the reason for its metric relations. In the case of a discrete manifold, the metric relations are contained in that structure, but in the case of a continuous manifold, they must come from outside. And he says "Either therefore the reality which underlies space must form a discrete manifoldness, or we must seek the ground ot its metric relations outside it, in binding forces which act upon it."[34] Deciding this question requires empirical investigations and the development of theories explaining their results. Theoretical groundwork as in this text is important for providing a general conceptual framework within which that can take place. But this leads into the realm of physics, which is outside the scope of this text.

References

Bao, David, Shiing-Shen Chern, and Zhongmin Shen. 2012. *An introduction to Riemann-Finsler geometry*. New York: Springer.

Blumenberg, Hans. 2007. *Die Genesis der kopernikanischen Welt*. Suhrkamp, Frankfurt/M.

Bottazzini, Umberto. 1999. Ricci and Levi-Civita: from differential invariants to general relativity. In *The symbolic universe: geometry and physics 1890–1930*, ed. Jeremy Gray. Oxford: Oxford Univ. Press.

Brouwer, Luitzen E.J.. 1911. Beweis der Invarianz der Dimensionszahl. *Mathematische Annalen* 11: 161–165.

Christoffel, Elwin Bruno. 1869. Ueber die Transformation der homogenen Differentialausdrücke zweiten Grades. *Journal für die Reine und Angewandte Mathematik* 70: 48–70.

Cogliati, Alberto. 2014. Riemann's commentatio mathematica, a reassessment. *Revue d'histoire des mathématiques* 20 (1): 73–94.

Darrigol, Olivier. 2015. The mystery of Riemann's curvature. *Historia Mathematica* 42 (1): 47–83.

De Risi, Vincenzo. 2007. *Geometry and monadology: Leibniz's analysis situs and philosophy of space*. Birkhäuser.

Di Scala, Antonio. 2001. On an assertion in Riemann's Habilitationsvortrag. *Enseignement Mathematique* 47 (1/2): 57–64.

[34] Let me also present the original German wording, because this is the key sentence when people discuss the vision of the unification of geometry and physics that Riemann may have had in mind: "Es muss also entweder das dem Raume zu Grunde liegende eine discrete Mannigfaltigkeit bilden, oder der Grund der Massverhältnisse ausserhalb, in darauf wirkenden bindenden Kräften, gesucht werden."

Ehresmann, Charles. 1948. Les connexions infinitésimales dans un espace fibré différentiable. *Sém. Bourbaki* 24: 153–168.

Ehresmann, Charles. 1951. Les connexions infinitésimales dans un espace fibré différentiable. *Colloque de Topologie, Bruxelles* 29: 55–75.

Eschenburg, Jost-Hinrich, and Jürgen Jost. 2013. *Differentialgeometrie und Minimalflächen*. Berlin, Heidelberg: Springer.

Finsler, Paul. 2013. *Über Kurven und Flächen in allgemeinen Räumen*. Berlin, Heidelberg: Springer.

Frobenius, Georg. 1877. Ueber das Pfaffsche Problem. *Journal für die Reine und Angewandte Mathematik* 82: 230–315.

Gauss, Carl Friedrich. 1973. *Werke*. Dieterichsche Universitätsdruckerei, Göttingen, 1870–1927. (Reprint Olms, Hildesheim, New York).

Gauß, Carl Friedrich. 1828. Disquisitiones generales circa superficies curvas. *Commentationes Societatis Gottingensis* 12: 99–146.

Gray, Jeremy. 2017. Riemann on geometry, physics, and philosophy—some remarks. In *From Riemann to differential geometry and relativity*, eds. Lizhen Ji, Athanase Papadopoulos, Sumio Yamada, 97–110. New York: Springer.

Hausdorff, Felix. 1914. *Grundzüge der Mengenlehre*. Leipzig: von Veit.

Hegel, Georg Friedrich Wilhelm. 1986. *Wissenschaft der Logik, I*. Suhrkamp: Frankfurt/M.

Jost, Jürgen. 2005. *Postmodern analysis*. New York: Springer.

Jost, Jürgen. 2006. *Compact Riemann surfaces. An introduction to contemporary mathematics*. New York: Springer.

Jost, Jürgen. 2017. *Riemannian geometry and geometric analysis*, 7th ed. New York: Springer.

Jost, Jürgen. 2025. *Bernhard Riemann: Ueber die Hypothesen, welche der Geometrie zu Grunde liegen*. Klassische Texte der Wissenschaft, Springer.

Jost, Jürgen. 2026. *Riemannian geometry and geometric analysis*, 8th ed. New York: Springer.

Jost, Jürgen, and Xianqing Li-Jost. 1998. *Calculus of variations*. Cambridge: Cambridge Univ. Press.

Kant, Immanuel. 1997. *Kritik der Reinen Vernunft (1781, 1787)*. (Hrsg. Jens Timmermann) Suhrkamp, Frankfurt/M.

Laugwitz, Detlef. 1996. *Bernhard Riemann 1826–1866*. Basel: Birkhäuser.

Lipschitz, Rudolf. 1869. Untersuchungen in Betreff der ganzen homogenen Functionen von n Differentialen. *Journal für die Reine und Angewandte Mathematik* 70: 71–102.

Pulte, Helmut. 2005. *Axiomatik und Empirie. Eine wissenschaftstheoriegeschichtliche Untersuchung zur Mathematischen Naturphilosophie von Newton bis Neumann*. Darmstadt: Wiss. Buchges.

Reich, Karin. 1997. *Die Entwicklung des Tensorkalküls. Vom absoluten Differentialkalkül zur Relativitätstheorie*. Birkhäuser, Basel

Riemann, Bernhard. 1892. *Gesammelte mathematische Werke und wissenschaftlicher Nachlass*. Bearbeitet von Heinrich Weber, Teubner.

Riemann, Bernhard. 1990. *Gesammelte mathematische Werke, wissenschaftlicher Nachlass und Nachträge. Collected papers*, ed. R. Narasimhan. Berlin: Springer.

Riemann, Bernhard. 2013. *Ueber die Hypothesen, welche der Geometrie zu Grunde liegen*. Hrsg. und kommentiert von Jürgen Jost, Klassische Texte der Wissenschaft, Springer.

Scholz, Erhard, ed. 2001. *Hermann Weyl's RAUM-ZEIT-MATERIE and a General Introduction to His Scientic Work*. Basel: Birkhäuser.

Scholz, Erhard. 1999. Weyl and the theory of connections. In *The symbolic universe. Geometry and physics 1890–1930*, ed. Jeremy Gray, 260–284. Oxford: Oxford Univ. Press.

Scholz, Erhard. 1982. Riemanns frühe Notizen zum Mannigfaltigkeitsbegriff und zu den Grundlagen der Geometrie. *Archive for History of Exact Sciences* 14: 213–232.

Shapiro, Stewart. 2000. *Philosophy of mathematics: Structure and ontology*. Oxford: Oxford University Press.

Sinaceur, Mohammed Allal. 1990. Dedekind et le programme de Riemann. *Revue d'histoire des sciences* 43: 221–296.

Spivak, Michael. 1975. *Differential geometry*. Berkeley: Publish or Perish.

Weyl, Hermann. 1993. *Raum, Zeit, Materie: Vorlesungen über allgemeine Relativitätstheorie*. Ed. Jürgen Ehlers. Springer.

Weyl, Hermann. 1913. *Die Idee der Riemannschen Fläche*. Leipzig, Berlin: Teubner.

Weyl, Hermann. 1918. Reine Infinitesimalgeometrie. *Mathematische Zeitschrift* 2: 384–411.

Weyl, Hermann. 1918. Gravitation und Elektrizität. *Sitzungsber. Preuss. Akad. Wiss.* 26: 465–480.

Reception and Influence of Riemann's Text 5

5.1 Helmholtz

For the understanding of Riemann's lecture and its importance, the comparison with the reasonings of the physiologist and physicist Hermann von Helmholtz[1] is particularly important.

[1] Hermann Helmholtz was born in 1821 as the son of a school teacher. For financial reasons, he initially had to work as a military surgeon, but had been able to study in Berlin with the leading anatomist and physiologist of his time, Johannes Müller (1801–1858). Having stepped forward with studies on the formation and propagation speed of nerve impulses and the paper *Über die Erhaltung der Kraft* (On the Conservation of Force, i.e., energy conservation), in 1849, he became professor of physiology in Königsberg, then in Bonn and Heidelberg. Among his significant achievements in sensory physiology were measuring the velocity of propagation of electrical nerve stimulations and the development of the ophthalmoscope. His monographs (Helmholtz 1856–1867) *Handbuch der Physiologischen Optik* (Handbook of Physiological Optics), Leipzig, Leopold Voss, in three installments from 1856 to 1867, and Helmholtz (1863) *Die Lehre von den Tonempfindungen als physiologische Grundlage der Musik,* (The Theory of Sensations of Tone as a Physiological Basis of Music), Braunschweig, Fr. Vieweg und Sohn in 1863, laid the foundations of systematic sensory physiology. The physiological research of Helmholtz and his colleague and friend Emil du Bois-Reymond (1818–1896) (brother of the mathematician Paul du Bois-Reymond (1831–1887)), the founder of electrophysiology and successor of Müller in Berlin, led to the final overcoming of the vitalist ideas, which their teacher Müller had still defended. Helmholtz' sensory-physiological investigations led him to an empiricist epistemology and on this basis to systematic considerations about the concept of space; these will be discussed in more detail in the text below. It is remarkable that the physiologist Helmholtz, who was only a mathematical autodidact, could penetrate so deeply into a basic question of mathematics, even if the details did not always withstand the professional criticism of the mathematician Sophus Lie (others, especially Klein [Bd. 1, S. 223–230]1926) (English translation in Klein (1979)), judged the contribution of Helmholtz significantly more generously than Lie, who could be unusually sharp also in disputations with other mathematicians whom he took as his competitors, like Killing or Klein). Helmholtz, who in the course of his career turned more and more to issues of physics, had in fact earlier

© The Author(s), under exclusive license to Springer Nature Switzerland AG 2025
J. Jost, *Bernhard Riemann — On the Hypotheses Which Lie at the Bases of Geometry*,
Classic Texts in the Sciences, https://doi.org/10.1007/978-3-031-91489-8_5

Helmholtz dealt in several journal articles and lectures with epistemological issues, in particular addressing the question of what we can learn about the structure of the world from our sensory experiences. His question thus was completely different from Riemann's natural philosophical question. Remarkably, his conclusions go first in the same direction as those of Riemann, but then take a different turn, because he makes a substantial additional assumption which he considers to be empirically evident, but which ultimately prevents him from reaching the generality of the theory of Riemann. Nonetheless this assumption turned out to be fruitful for the development of mathematics because it provided a major impulse for Lie's theory of transformation groups, which together with Riemannian geometry became fundamental for modern physics. In fact, the thrust of Helmholtz' arguments was directed against Kant's philosophy of space as a synthetic a-priori construction, rather than against Riemann's theory.

We refer here to Helmholtz's writings *Über den Ursprung und die Bedeutung der geometrischen Axiome* (On the Origin and Importance of geometrical axioms), *Über die tatsächlichen Grundlagen der Geometrie* (About the factual basis of Geometry), *Über die Tatsachen, die der Geometrie zugrunde liegen* (On the facts on which geometry rests), that article which most clearly relates to Riemann and already in its title (which replaces Riemann's "hypotheses" by "facts") seems to contain a criticism against him, and finally *Die Tatsachen in der Wahrnehmung* (The facts in Perception), together with their three supplements.[2] (To elucidate the epistemological position of Helmholtz, also his later article *Zählen und Messen, erkenntnistheoretisch betrachtet* (Counting and measuring, epistemologically considered) is useful. There Helmholtz, incidentally, shows himself much more conciliatory

obtained an important and difficult mathematical result in hydrodynamics. He proved that vortices are conserved in a frictionless fluid. For that work, incidentally, Riemann's theory of conformal mappings had been an important inspiration. His work, and that of his student Heinrich Hertz (1857–1894), contributed decisively to a general acceptance of the Faraday-Maxwell theory of electrodynamics. Helmholtz' approach to derive the electrodynamic field equations from a principle of least action was an important precursor for development of the theory of relativity. Helmholtz's own theoretical approach led to the prediction of the existence of the electron, but ultimately proved to be futile, because it was based on the existence of the ether. In 1871, Helmholtz became professor of physics in Berlin. He was ennobled in 1883 (and his family name was changed into *von* Helmholtz as part of this procedure). In 1888, he was appointed president of the newly founded Physikalisch-Technische Reichsanstalt (Physico-Technical State Institute), a pioneering large scale research institution both through its research agenda and its organizational principles. Helmholtz died in 1894. Helmholtz was the great universal scientist of the second half of the nineteenth century, and he also enjoyed the corresponding social recognition and prestige. His position in German science can perhaps be compared with that of Alexander von Humboldt in the first half of the nineteenth century. For his biography and scientific role and achievements, see Koenigsberger (1903). A more recent study is Schiemann (1997). There exists an extensive literature on Helmholtz. I mention only the recent work of Meulders (2010).

[2] For references see the bibliography at the end. In the sequel, I shall cite these references in abbreviated form as *Axiome*, *Grundlagen*, *Geometrie* and *Wahrnehmung*, the first and the last and also the commentaries by Hertz and Schlick with the page numbers of the edition of F. Bonk (Helmholtz 1998), the others from *Wissenschaftliche Abhandlungen*, Vol. II (Helmholtz 1883).

towards Kant, by accepting the basic idea of space as a transcendental form of intuition and only attacks a special position which according to him, is an unfortunate later addition by Kant's followers.) We treat the documents mentioned here as a unit, even if in the course of time, the thinking of Helmholtz certainly evolved. In particular, at the beginning he was not yet aware of the possibility of non-Euclidean (hyperbolic) geometry.[3]

The fundamental problem (*Geometrie*, p. 618) that Helmholtz poses is that of a distinction between the objective content of geometry and that part that is or may be set by definitions or that depends on the form of presentation, for example the choice of coordinates, and that is consequently not invariant. Helmholtz aimed primarily against the idea of space that Kant had developed in his critical writings and that we have outlined above, namely that space is an a priori given form of all external intuition.[4] Helmholtz works out the difference between a purely formal scheme, "in which any content of experience would fit",[5] and one whose perceptible content is limited or constrained from the outset. The first one he can accept, the second, however, he rejects. He agrees with Kant that the general form of spatial intuition is transcendentally given. For him, this ultimately means that space is a continuous manifold that makes the coexistence of different bodies possible, and thus their juxtaposition,[6] and in which magnitudes can be compared. However, more detailed determinations have to be taken from experience, instead of being given before all possible experience.[7] Helmholtz begins his argumentation in *Axiome* with the axioms of Euclidean geometry. Axioms cannot be proved, and he therefore raises the question of why we nevertheless accept these axioms as correct. (Hilbert will elaborate later that axioms are arbitrary postulates, which in some sense eliminates the reason for Helmholtz' question.) In his response he is guided by the fundamental proof scheme of Euclidean geometry, the demonstration of the congruence of two- or three-dimensional geometric figures. This is based on the assumption that geometric objects can be moved freely in space without changing their shape. That, however, and this constitutes the central point of Helmholtz' argument, is not a logical necessity, but an empirical fact.[8]

---

[3] e.g. *Grundlagen*, pp. 613, 615. This is corrected only in the supplement to this article. likewise *Geometrie*, pp. 637–9, where it is corrected in footnotes inserted in *Wissenschaftliche Abhandlungen*.

[4] To what extent Helmholtz has misunderstood the Kantian notion of synthetic a priori judgment by not recognizing the difference between logical and descriptive necessity was indeed an essential aspect of the argumentation of the Kantians, but may be left open here. See also the remarks of Schlick, p. 49.

[5] "in welches jeder beliebige Inhalt der Erfahrung passen würde", in *Axiome*, P. 16.

[6] Concerning this issue, modern mathematics has then gone even further in the direction taken by Helmholtz, insofar as also topological and not only metric properties of space may be contingent.

[7] *Wahrnehmung*, p. 159.

[8] Apparently Helmholtz was not aware that it had already been an essential postulate of Leibniz that every body needs to be thought of as movable in space without change of form, see pp. 161, 168 in Volume V of Gerhardt (1860). This is constitutive for Leibniz's constructive approach in his geometry of position, see Cassirer (1902). For Leibniz, however, what seemed clear to Helmholtz as an empirical fact, was still a mathematical and philosophical problem (De Risi 2007). Leibniz

What we can imagine is limited by the structure of our sensory organs, which are adapted to the space in which we live. More precisely, we construct the space from the data on our two-dimensional retina. First, this provides a new empirical turn to the old philosophical argument of Leibniz for the relativity of space, namely that it is not possible to determine if all objects are moved or enlarged in the same manner, because such a change would also affect our sensory organs. Second, this reconstruction is flexible to some extent. A person who puts glasses in front of his or her eyes that convexify everything so that he or she sees objects as they would appear in the hyperbolic space,[9] after a short while will adapt himself or herself to this new visual experience and have no problems orienting himself or herself in space. Likewise, we could also become used to living in a non-Euclidean geometry. What is important is the internal consistency of the perception of space, as long as no other physical phenomena come into play. (A well-known example is an experiment with reversing eyeglasses. A person who is given such reversal glasses, which have the effect of reversing top and bottom, so that everything seems to stand on its head, will get used to it after a while and then find his or her bearings in the world again without any problem. In particular, all movements and actions are matched to what is seen through the reversing glasses. When taking off the reversal glasses, the test person needs some time again to get familiar with the world, until the objects cease to seem to stand on the head.) For Helmholtz, it is therefore crucial that our perception of space be constructed from perceptions and sensations that are consistent among each other and in themselves. To the physiologist Helmholtz we owe in fact the fundamental insight that the brain constructs an image of the outside world out of local electrical activities, which are propagated with a measurable, finite rate along nerves (*"The sensations are for our consciousness signs, and to learn to understand their meaning is left to our intellect"* or "Insofar as the quality of our sensation is a message for us of the peculiarity of the external influence, by which it is excited, it can be taken as a *sign* of the same latter, but not as an *image*. ... A sign need not have any kind of similarity to what it stands for. The relationship between the two is limited to that the

carefully analyzed the difference between similarity and congruence of geometric figures. Without a direct comparison with respect to a common scale, one can determine only the similarity, i.e. the equality of the internal relations of two figures, but not their congruence, i.e. the absolute equality of their magnitudes. Leibniz does not argue with the mobility of the rigid scale, but with that of the figures to be examined, which of course also leads to the homogeneity of space. Kant is also familiar with these issues. One could now, casually speaking, think that the physicist walking around with a yardstick in the field simply ignores a pseudo-problem of the mathematician struggling with the penetration of Euclidean geometry or of the philosopher speculating in his study room. However, the situation is not that simple. As will be explained in Sect. 5.4, Weyl later proposed to allow even a path-dependent gauge freedom in the measurement units, so that lengths can change when a body is transported in space. This idea was ultimately rejected by physicists, for example by reference to the absolute length scale of atomistics. But as explained in Sect. 5.4, this idea had become central for modern elementary particle physics in a somewhat different way.

[9] Helmholtz shows at this point a deep understanding of the geometric model of non-Euclidean space by Beltrami cited below (at that time and also by Helmholtz called pseudospherical geometry).

same object, acting under the same circumstances, causes the same sign")[10], and this then leads into modern constructivism as an approach to philosophy building on neurobiological insights.[11] However, according to Helmholtz, the law of causality has to be presupposed for the interpretation of our experiences.[12] Experiences thus are not arbitrary, but refer to an external world, the physics and geometry of which is to be reconstructed.

However, such an adaptation to the geometric relations of the outside world has specific limitations caused by the structure of our sensory organs. In particular, this concerns the dimension of space.

To illustrate this, Helmholtz invokes the conceptual model of rational beings that live on a surface, i.e., in a two-dimensional world, and therefore cannot imagine a third dimension.[13] A way out for the flatlanders, which want to conceive the third dimension, as we like to think of the fourth dimension, is offered by the formal computational methods of mathematics that can perform constructions in any dimensions without constraints. The measurements in the empirically given space can then also be compared with the results of calculations in coordinate systems in constructed spaces, and in this way the special properties of empirical space can then be identified. This is what Helmholtz considers as Riemann's approach. In particular, according to Helmholtz, the empirical space is not a general three-dimensional manifold in Riemann's sense, but determined by additional properties, firstly, of the free mobility of bodies without change in shape to all points and in all directions[14] and second, the vanishing of the curvature. In fact, from the free mobility of bodies there follows, first, the infinitesimal validity of the Theorem of Pythagoras, as assumed by Riemann, and this is a major mathematical contribution of Helmholtz. Second, there even already follows the constancy of the curvature, and this is also a important mathematical result (although this

---

[10] *"Die Sinnesempfindungen sind für unser Bewußtsein Zeichen, deren Bedeutung verstehen zu lernen unserem Verstande überlassen ist"*, in Helmholtz (1910, S. 433) (emphasis in the original) or "Insofern die Qualität unserer Empfindung uns von der Eigentümlichkeit der äußeren Einwirkung, durch welche sie erregt ist, eine Nachricht gibt, kann sie als ein *Zeichen* derselben gelten, aber nicht als ein *Abbild*. ... Ein Zeichen aber braucht gar keine Art der Ähnlichkeit mit dem zu haben, dessen Zeichen es ist. Die Beziehung zwischen beiden beschränkt sich darauf, daß das gleiche Objekt, unter gleichen Umständen zur Einwirkung kommend, das gleiche Zeichen hervorruft", in *Wahrnehmung*, p. 153 (emphasis in the original).

[11] On the history of neuroscience, see Breidbach (1997). Here, we cannot discuss the development of sensory physiology before, by, and after Helmholtz, or the influence of Lotze's theory of local signs ("Lokalzeichen") or the dispute between empiricists like Helmholtz and nativists like Hering (for Helmholtz' position, see e.g. *Wahrnehmung,* p. 163f.) and other such issues.

[12] *Wahrnehmungen*, p. 171f, p. 191.

[13] This idea was later elaborated and poopularized by Abbott (1884) which he published under the pseudonym A. Square. Actually, before Helmholtz, this idea had already been mentioned by Gauss, see von Waltershausen (1856, p. 81). But even before Gauss, the founder of psychophysics, Gustav Theodor Fechner (1801–1887), had proposed a similar idea, s. Thiele (2003).

[14] Helmholtz also works out the monodromy principle that a body after a rotation by 360° again returns to its original position and shape. Lie then later criticized that this is not an independent axiom, as Helmholtz believed , but that it follows from the other axioms of Helmholtz.

had already, as stated above, been found by Riemann, but Helmholtz based his analysis on a different set of axioms than Riemann, so that Helmholtz' results do not directly follow from those of Riemann), even if Lie later criticized the stringency of Helmholtz' mathematical deductions. That the curvature must be constant, thus follows, according to Helmholtz, already from a general principle of experience, while the exact value of the curvature then is the result of a specific empirical measurement.[15]

Helmholtz also notes that the free mobility of bodies is not a purely geometric property. Namely, if all bodies changed in the same way when changing location, we would have no way to determine this, because also all measuring devices would change with the bodies. So here an additional physical principle is needed again. This is a subtle point, however. For what is rigid can ultimately neither be derived from principles nor determined empirically. In order to verify that a body is rigid, we would need to verify that the distances between the individual points in this body do not change, but for that purpose we would need again a rod that has already been verified as rigid. According to Einstein, the determination of what is rigid can therefore be based only on a convention. The physical principle can then only be the simplicity of explanation,[16,17] while Helmholtz still believes that he can use the physical behavior of inertial bodies. At this point perhaps also the main difference from the approach of Riemann becomes clear. Helmholtz' reasoning depends on the assumption of the existence of rigid bodies while Riemann assumes only consistent length scales. The physical principle of rigid bodies, which Helmholtz introduces, prevented him from reaching the principle of general relativity, where the behavior of bodies and the geometry of space become intertwined. For Riemann, the metric field of space is not necessarily rigid, but can interact with the matter located in space. As Riemann has set up the theory, it is in particular possible that a body carries the metric field along. The metric field is then determined or can be altered by the body during its motion. In this way, the motion of rigid bodies becomes possible also in an inhomogeneous geometry. The geometry would thus become time-dependent, again a central point in Einstein's theory. For Helmholtz, however, this entanglement of the geometry of space and the behavior of bodies instead takes place only in perception. Although it is a consequence of the empiricist assumptions of Helmholtz

[15] This again is part of a long line of discussion. That the curvature of space must be empirically measurable, was already known to Gauss. Whether the curvature of space actually vanishes on a cosmic scale, leads into the even today still ongoing discussion about the cosmological constant of Einstein, which recently has been revived. This came about because of some phenomena that cannot be explained by established cosmological physics. This leads to the search for so-called dark matter and dark energy.

[16] Concerning this issue, see the explanations of Schlick, p. 52.

[17] This is the philosophical direction of conventionalism (see below). Carrier (1994) concludes that Helmholtz has thus stimulated several different directions of the philosophy of physical geometry because his views both can be interpreted in such a way that both the free mobility of rigid bodies is an empirical fact, and that it provides a useful convention, and finally, that it is the precondition of physical and geometrical measurements. A detailed description of the history of ideas of the arguments of conventionalism is found in Carrier (2009).

that he admits that the spatial independence of the mechanical and physical properties of bodies could in principle also be refuted by experience, he does not seem to have seriously considered that this assumption of position independence might actually be empirically false. His concern is rather the anti-Kantian argument that the perception of objects and their spatial relationships is derived empirically and not given prior to all experience.

In his *Geometrie* , Helmholtz then deduced from four axioms, which he had already outlined in his *Grundlagen*, that a space satisfying those axioms, which are compatible with the empirical intuition, necessarily is a space of constant curvature in the Riemannian sense. These axioms are (see *Grundlagen*, p. 614f.)

1. Specified dimension n and representability by coordinates that change continuously under the continuous motion of a point (in Riemann's terminology, this just means that space is an n-dimensional manifold).
2. Existence of bodies that can move and are rigid in the sense that the distances between any two of their points remain invariant.
3. Free mobility: bodies can move as a whole (but not within themselves, that is, internally), i.e. motions are only constrained by the invariances of the interior distances as postulated in (2), and congruence between two bodies does not depend on their position in space.
4. Monodromy: The full rotation about an axis brings a body back to itself.

On the necessity and independence of these axioms see the investigations of Lie (1890) and the comments of Hertz in Helmholtz (1883). The mathematical deductions of Helmholtz, which incidentally are restricted to the case of dimension 3, are no longer of interest for us for the reasons already set out.

Even apart from the fact that the physical assumption of rigid bodies turned out to be an obstacle and unfortunate in the light of the subsequent development of physics, the Helmholtz approach, already in the explicit opposition between "facts" versus "hypotheses" in the respective titles, takes back that aspect of Riemann's approach that was pioneering not only for the development of mathematics. This is the investigation of ideal "spaces", in the sense that they are freely constructed by our imagination, instead of only empirically given ones. And as will be explained below, this conceptual step of Riemann then also opens up fundamentally new perspectives for physics.

Subsequently, the strands from mathematics, physics, philosophy and sensory physiology that came together through the work of Riemann and Helmholtz parted again. Therefore, the reception history consists of multiple parallel strands, often even within the participating sciences. In the following, we shall try to describe and analyze some of them. We shall, however, not discuss in detail the many objections to the considerations of Riemann and Helmholtz that are typically based on misunderstandings or false reasonings,[18] even if they were prominent in the discussion of their times and thus in a certain sense also important for the reception history. In fact, since Helmholtz had directly challenged the philosophers

[18] Many examples are presented and analyzed in Torretti (2012).

of Kantian obedience, diverse criticisms were raised not only against him, but also against Riemann's considerations, which Helmholtz had invoked. One of the first critics was the Göttingen philosopher Hermann Lotze (1817–1881), who, although he had probably been already present at Riemann's habilitation lecture, had recognized its importance apparently only from the philosophical turn which Helmholtz had given to the discussion, and was then alarmed by it. In particular, Lotze rejected the relation, postulated by Riemann and Helmholtz, between space and the physical processes taking place in it, and therefore also the possibility of an empirical examination of the properties of space. His argument was that, even if the behavior of physical bodies at the astronomical scale should show a deviation from Euclidean geometry, instead of abandoning our idea of a Euclidean space, we rather should adopt a new physical force which causes a deviation of the propagation of light rays from the Euclidean straight line. This argument continued to play a role in the conventionalism of Henri Poincaré (see below). Lotze's attempts to engage in a real mathematical reasoning appear clumsy, however. Likewise, the arguments of other critics, such as the psychologist Wilhelm Wundt (1832–1920) or the French neo-Kantian Charles Renouvier (1815–1903), were found to be unsound. Also the philosopher Bertrand Russell (1873–1970) later dealt with little success and some fallacies with the matter.[19]

So much for a brief sketch of the discussions at that time. One might nevertheless suspect that, despite the rather hopeless attempts of the contemporary philosophers and followers of Kant, the situation today, one and a half centuries later, perhaps looks different and Kant ultimately could still have gained the upper hand against Helmholtz. After all, Kant has survived the course of time significantly better than Helmholtz. Kant is (still or again) recognized as one of the greatest if not the greatest philosopher of modern times. Helmholtz is considered as an eminent physicist, indeed, but today, he is viewed more as a figure of transition, whose contributions shaped the future of physics less than those of Maxwell or Boltzmann. Maxwell's theory of electromagnetism remains valid today and constituted an essential basis of Einstein's special theory of relativity, Boltzmann's reflections on statistical physics initiated an entirely new way of thinking which remains highly significant for current physics.

Helmholtz' importance shrinks further when compared to the great achievements of twentieth century physics, Einstein's theory of relativity and quantum physics, which is connected with the names of Planck, Bohr, Heisenberg and Schrödinger. Even theorists of neurophysiology, the field that Helmholtz founded and profoundly shaped both experimentally and conceptually through his considerations and experiments on the processing of sensory stimuli and on the representation of the external world in the nervous system of the brain, today talk more often about Kant than about Helmholtz. How can Helmholtz's critique of Kant then be evaluated today? For this purpose, we must draw upon both what has been set forth already and what remains still to be discussed, but perhaps such a combination of retrospection and anticipation at this point might be useful for the understanding of the situation of the problem and the historical context of the old discussion. Newton's theory of

[19] see Russell (1897, 1899) and the penetrating analysis of Torretti (2012).

gravitation, prepared by Kepler, proposed the interaction of bodies in space. This was much more than just a juxtaposition of objects, such as in the theories of Aristotle and Descartes. By his ontology of absolute space, Newton prevented himself from developing the radical explosive power of this concept. Newton's antagonist Leibniz shifted attention to the consistency of the relations of objects with each other, but did not have an appropriate notion of physical force in order to translate this into a physical theory.[20] Kant then examined the preconditions in the perceiving subject for perception of such coherent relations, but again consisting only in a juxtaposition and not in an actual causal relationship. (For Kant, the force of attraction belonged to the realm of dynamics, which in contrast to geometry, which is synthetic a priori, depended on empirical perceptions.) Helmholtz objected against this that these conditions not only lie in the perceiving subject, but have to be determined by physical measurements. Here, physics wants to wrestle a piece of reality out of the hands of philosophy. But this piece of physics is only about position, but not about causal interactions, and those were even made impossible by founding it on the assumption of rigid bodies. In this sense, Helmholtz' objection is indeed justified, but does not penetrate to the actual physical heart of the matter.

Riemann is not well known outside the circles of mathematicians, but remains recognized within mathematics as one of the greatest even today without any signs of weakening. He had an approach that was motivated by questions of natural philosophy, but was more general and mathematically and structurally conceived. His approach laid the basis that the gravitational effects of bodies upon each other couuld be modelled geometrically in the theory of general relativity. This poses difficulties for both Kant and Helmholtz, and perhaps this also is the answer to our question.

5.2 The Further Development of Riemannian Geometry and Einstein's Theory of Relativity

5.2.1 The Impact of the Theory of Relativity

In mathematics, Riemann's geometric considerations were taken up by Elwin Bruno Christoffel (1829–1900) and Rudolf Lipschitz (1832–1903) in Germany and by Eugenio Beltrami (1835–1900) and Gregorio Ricci-Curbastro (1853–1925) in Italy. Beltrami, who had already developed geometric realizations of the non-Euclidean geometry of Bolyai and Lobatschewsky (Beltrami 1868a, c), then also was the first to identify non-Euclidean geometry as a special case of the general Riemannian geometry (Beltrami 1868b). Felix Klein

[20] Leibniz had the concept of *vis viva*, living force, essentially today's energy, and demonstrated that it is preserved in mechanical interactions. This was an important result for the future of physics. The vis viva concept, however, does not account for the interactions between different bodies in space without direct physical contact. Nevertheless, Leibniz's concept of *action* lead to the principle of least action, a starting point for many physical theories, including ones going far beyond Newton. It is still disputed among historians to what extent that principle was already clear to Leibniz, however.

(1849–1925) embedded these geometries in a comprehensive geometric program.[21] However, the general geometry conceived by Klein, projective geometry, is different from that of Riemann. In contrast to Riemannian geometry, it is not based on lengths and distances, but rather on proportions. Although the projective space also carries a Riemannian metric, for Klein, the transformation properties instead of the metric relations were fundamental. Klein's approach, which was very influential at the time,[22] may have delayed the intensive reception of the ideas of Riemann in Germany. Nowadays, however, the approaches of Riemann and Klein are no longer seen as incompatible or competing with each other.[23]

Ricci and Levi-Civita developed the tensor calculus of Riemannian geometry in the form that is essentially still used today. This tensor calculus then formed the mathematical basis of the general theory of relativity of Einstein. Subsequently, Riemannian geometry has been further developed by Élie Cartan (1869–1951) and Hermann Weyl (1885–1955) and many others and experienced a great boost and a momentum that is unbroken until today.[24] The considerations of Weyl on infinitesimal geometry and the concept of an affine connection have already been mentioned in the context of Riemann's lecture, see p. 99. In particular, his book *Space, Time, Matter*[25] has been very influential for the mathematical and conceptual foundations of general relativity. The development of Einstein's theory itself will not be detailed here, because the basic papers of Einstein have been published and commented in another volume in this series,[26] as is Weyl's *Space, Time, Matter*.[27] In Einstein's theory, the geometry of space-time is shaped by the gravitational effects of the masses contained therein. In Newtonian physics, the inertial mass of a body originated in its resistance to changes in motion, while its gravitational mass expressed the response to the attractive forces of other bodies. Why the two were always proportional and thus by suitable normalization could be equated with each other, the theory could not explain. In Einstein's theory, however, the two concepts coincide. Einstein realized that the effects of acceleration and gravity cannot be distinguished. Both inertial and gravitational mass are derived from the resistance

---

[21] Klein (1974), with additions published in Klein (1972); also Klein (1871) and the posthumous monograph (Klein 1928) and his collected works (Klein 1921–1923). An English translation with an extensive historical commentary (Rowe 2025) has also appeared in this series. On the programs of Lie and Klein see also Hawkins (2012, 1984). For the further development of Klein's program e.g. Sharpe (2000).

[22] Hawkins (1984), however, comes to the conclusion that Klein's manifest itself actually did not exert a significant influence, but that programmatically related ideas were developed more or less independently not only by Lie, but also by Eduard Study, Wilhelm Killing and Henri Poincaré. Anyway, like Klein's, neither of these approaches granted a fundamental role to Riemannian metrics.

[23] This is due in particular to the geometer Élie Cartan, see below, p. 139.

[24] For a presentation of the current state, see for instance (Jost 2017).

[25] For an edition with a detailed commentary, see Weyl (2025).

[26] Michel Janssen und Jürgen Renn (2022).

[27] Weyl (2025), with a detailed commentary by Domenico Giulini and Erhard Scholz, based on the 5th German edition of 1923. The only available English translation (Weyl 1952) of Weyl's text is not very satisfactory, unfortunately.

to changes of motion. The reference motions, however, with respect to which a change is relevant, no longer are the uniformly accelerated motions in a space that is conceived as Euclidean, absolute and therefore also independent of the masses located in it. Those reference motions rather are the movements along geodesic paths in a certain Riemannian (more precisely, Lorentzian) space-time under the influence of the gravitational effects of the masses contained in it. Gravity thus does not act instantaneously and unmediated through empty, absolute space on distant bodies, but locally shapes the geometry of space-time, which in turn determines the motion of bodies. In short, in Newtonian physics, bodies move under the influence of gravitational forces on curved paths in a linear (i.e. non-curved, Euclidean) space. In Einstein's physics, however, they move on straight lines (i.e. geodesic curves) in a curved space. Gravity no longer bends the trajectories of bodies, but the space in which they move. Einstein's field equations couple the Riemannian curvature of space-time with the energy-momentum tensor of matter. The presence of matter thus changes the geometry of space-time, and acceleration is measured now in relation to this Riemannian geometry instead of an independent absolute Euclidean one. The Einstein field equations are themselves derived from symmetry principles, specifically from the requirement of general covariance. The physical laws should apply regardless of the chosen coordinates and therefore the field and motion equations expressed in coordinates must transform under coordinate changes suitably, i.e. obeying specific rules.

5.2.2 The General Theory of Relativity and the Problem of Space

During the development of the general theory of relativity, Einstein created an obstacle for himself with his "hole argument". He believed he had found an argument against a fully covariant theory. Imagine removing a small compact region from spacetime. In other words, make a hole in spacetime. In spacetime, a solution to the field equations exists, both within the hole and on its boundary. Now, let spacetime undergo a diffeomorphism φ,[28] which is the identity outside the hole, fixing all points of spacetime there, particularly on the boundary of the hole. However, inside the hole, it changes the positions of the points. Under general covariance, i.e., independence under diffeomorphisms, φ transforms the solution of the field equations into another solution. This contradicts the physical principle that a solution should be uniquely determined by its boundary conditions at the edge of the hole (or, as the development of the theory shows, by the conditions on a Cauchy hypersurface in the complement of the hole, see Sect. 5.2.3).

As long as diffeomorphisms are understood only as coordinate transformations, acting in a passive sense, there is no problem. But if they play an active role and shift the points of an underlying spacetime, this seems to lead to a non-uniqueness of the solutions to the field equations.

[28] A diffeomorphism is a mapping that is invertible and differentiable, along with its inverse.

It was only after some time that Einstein realized this was a pseudo-problem. His later discussion with Moritz Schlick[29] significantly contributed to clarifying the issue. The outcome of this discussion was that spacetime points do not have independent existence. Therefore, solutions transformed by a diffeomorphism are not physically distinct. According to this interpretation, there is no independently underlying spacetime whose points can then be equipped with a metric tensor. Instead, points are defined by a field, in this case, the metric tensor. Thus, points are not elements of spacetime but intersections of world lines.[30]

This also sheds deeper light on the requirement of general covariance. It concerns not only the principled indistinguishability and equivalence of different descriptions. Rather, the old Leibnizian idea reemerges, that the homogeneity of space implies a formlessness that leads to the indifference of its parts or elements to one another. As a result, there is no rational justification for specific configurations in space (or time). Without an assignment of physical attributes, points in space cannot be rationally distinguished from one another.[31] This was likely also Riemann's idea, that a manifold describes a general concept that allows various determinations. Any physical theory must, in principle, be independent of the description of the underlying objects, provided these descriptions capture the same aspects and merely represent them in different coordinate systems. However, the central task of a physical theory is to clarify through which physical properties these objects can be distinguished from one another. In Riemann's concept of a manifold, both aspects are inherent: that the same point in the manifold can be described and represented differently in coordinates, and that in the manifold, as long as no additional structure is introduced, all points are of the same type and can be transformed into one another within the manifold (homeomorphisms in mathematical terminology). The concept of a manifold thus captures the multitude of points but does not yet provide a criterion for their identification or differentiation. Metric relations then provide distinguishing relationships between points, and curvature values can assign specific characteristics to individual points. As Riemann recognized, geometry cannot be derived from the concept of a manifold alone but requires a physical determination. Einstein's theory accomplishes this systematically and fundamentally. In quantum theory, however, according to Heisenberg, this aspect is reversed. Here, the uniform or identical object appears only in different manifestations. Physically accessible are only these manifestations, not the object itself.

Precisely this coordinate independence of geometric relationships and physical laws was one of the central ideas in Riemann's theory, and it found its formal expression in tensor calculus (see Sect. 4.4.2), which was developed and refined by Riemann's successors. This made

[29] See, for example, Engler and Renn (2018).

[30] After formulating the field equations of the general theory of relativity, Einstein did not revisit the hole argument. It was much later brought back to light by Stachel (1989), and following a paper by Earman and Norton (1987), it sparked an intense philosophical discussion, marked by numerous misunderstandings, which I will not delve into here.

[31] In that direction, Minkowski writes "In order to leave nowhere a gaping void, we want to imagine that everywhere and at any time there is something perceptible." (Lorentz et al. (1923, p. 55), my translation).

Riemannian geometry so useful for Einstein. It was crucial, of course, that the Riemannian formalism could naturally be extended from spatial problems to spacetime problems, despite the fundamental difference that the metric tensor is no longer positive in all directions, but spatial and temporal directions acquire opposite signs. These corresponding structures are then called Lorentzian (instead of Riemannian) manifolds. The reference space here is no longer Euclidean, but Minkowski space.[32] We will now explain this mathematically.

5.2.3 The Mathematical Formalism of General Relativity

To develop the formalism, we first need to introduce the concept of the *Lorentz metric*. Although this concept is mathematically meaningful in any dimension, we restrict ourselves here to the four-dimensional spacetime. As in (4.38), we consider a symmetric tensor of the form

$$g_{\mu\nu}(x)\mathrm{d}x^{\mu}\mathrm{d}x^{\nu} \tag{5.1}$$

where $g_{\mu\nu} = g_{\nu\mu}$, and the indices μ, ν range from 0 to 3. The index 0 represents the time direction, while 1, 2, and 3 represent spatial directions. In matrix form:

$$\begin{pmatrix} g_{00} & g_{01} & g_{02} & g_{03} \\ g_{10} & g_{11} & g_{12} & g_{13} \\ g_{20} & g_{21} & g_{22} & g_{23} \\ g_{30} & g_{31} & g_{32} & g_{33} \end{pmatrix} \tag{5.2}$$

Such a Lorentz tensor is no longer positive definite but has the signature $+---$. When diagonalized via a linear transformation such that all $g_{\mu\nu}$ with $\nu \neq \mu$ vanish, we find that $g_{00} > 0$, while $g_{11}, g_{22}, g_{33} < 0$.[33] Like a Riemannian metric, a Lorentz metric $g_{\mu\nu}(x)$ generally depends on the point x. The simplest case is the *Minkowski metric* with constant values:

$$\begin{pmatrix} c^2 & 0 & 0 & 0 \\ 0 & -1 & 0 & 0 \\ 0 & 0 & -1 & 0 \\ 0 & 0 & 0 & -1 \end{pmatrix} \tag{5.3}$$

where c is the speed of light. (Often, units are chosen such that $c = 1$, as c is merely a convention relating space and time measurements.) The four-dimensional space equipped with the Minkowski metric is called *Minkowski space*.

Since a Lorentz metric g is no longer positive definite, there exist directions $V = v^{\mu}\frac{\partial}{\partial x^{\mu}}$ satisfying

---

[32] Minkowski (1909a, b); reprinted, for example, in Gauss et al. (2013, 100–113). For English translations of Minkowski's works, see Minkowski (2013).

[33] In the literature, the opposite convention is often used, where $g_{00} < 0$ and $g_{11}, g_{22}, g_{33} > 0$. This makes no physical difference.

$$g(V, V) = g_{\mu\nu} v^\mu v^\nu = 0. \tag{5.4}$$

Physically, these correspond to the directions of *light rays*. In the Minkowski metric (5.3), these rays travel in straight lines at the speed of light. Directions V with $g(V, V) > 0$ are called *timelike*, and those with $g(V, V) < 0$ are called *spacelike*.

Even though a Lorentz metric is no longer positive definite, the formal constructions previously applied to a *Riemannian metric* can also be performed for a Lorentz metric. Specifically, we can define a curvature tensor as in Definition 4.2:

$$\begin{aligned} R^q_{ikj} &:= \frac{\partial \Gamma^q_{ij}}{\partial x^k} - \frac{\partial \Gamma^q_{ik}}{\partial x^j} + \Gamma^p_{ij}\Gamma^q_{pk} - \Gamma^p_{ik}\Gamma^q_{pj} \\ \text{and } R_{\ell ijk} &:= g_{\ell q} R^q_{ijk}, \end{aligned} \tag{5.5}$$

where the Christoffel symbols are defined as before.

In Riemannian geometry, additional curvature quantities can be derived from the curvature tensor. Although not explicitly detailed in Riemann's work, these concepts are required for the formulation of Einstein's field equations. Since these are general constructions, we again work in a general dimension n, using Latin indices instead of Greek ones.

First, the Ricci tensor is defined with components:

$$R_{ij} = g^{k\ell} R_{ikj\ell} = R^k_{ikj}, \tag{5.6}$$

a contraction of the curvature tensor. The Ricci tensor is symmetric:

$$R_{ij} = R_{ji} \tag{5.7}$$

for all indices. It thus has $\frac{n(n+1)}{2}$ degrees of freedom, but n of these are due to coordinate choices. Consequently, the Ricci tensor has as many true degrees of freedom as the metric tensor. By Riemann's heuristic counting argument, the Ricci curvatures should determine the metric. Although this is subtler in practice, the Ricci tensor appears in Einstein's field equations to determine the spacetime metric.

n of the degrees of freedom of the Ricci tensors disappear because from (4.192) we get the contracted second Bianchi identity

$$2\nabla_{\frac{\partial}{\partial x^j}} R^j_i = \nabla_{\frac{\partial}{\partial x^i}} R\,, \tag{5.8}$$

valid for all indices $i = 1, \ldots, n$. The second Bianchi identity is again an expression or a consequence of the diffeomorphism invariance of the metric. The general principle behind was discovered by Noether (1918).

Another contraction yields the *scalar curvature*:

$$R = g^{ij} R_{ij}. \tag{5.9}$$

Now, the *Einstein field equations*, the fundamental equations of *General Relativity*, can be expressed as:

$$R_{\mu\nu} - \frac{1}{2} g_{\mu\nu} R = \frac{8\pi G}{c^4} T_{\mu\nu}. \tag{5.10}$$

Here, $T_{\mu\nu}$ is the *energy-momentum tensor* (also called the *stress-energy tensor*), which describes the distribution of matter, radiation and non-gravitational forces, and G is Newton's gravitational constant. By (5.8), the left-hand side of (5.10) is divergence-free, consistent with the divergence-free nature of the stress-energy tensor.

In vacuum, i.e., in the absence of matter, the Einstein field equations reduce to:

$$R_{\mu\nu} - \frac{1}{2} g_{\mu\nu} R = 0. \tag{5.11}$$

The Einstein field equations in a vacuum do not only have the Minkowski metric as a solution but also many others, including some with singular or peculiar spacetime geometries. In particular, it does not follow from (5.11) that the curvature tensor vanishes.

Since, due to the $+ - --$ signature of the metric in (5.11), it is a hyperbolic differential equation for the metric, initial conditions are needed to determine a solution.[34] We will briefly outline this in the case of a vacuum, i.e., for (5.11). The process begins with what is known as a Cauchy hypersurface. This is a sufficiently smooth three-dimensional spacelike submanifold of spacetime that is complete in the sense that every timelike curve extending infinitely into both the past and the future intersects it at least once. Moreover, such an intersection must occur exactly once to exclude closed timelike curves. To embed such a Cauchy hypersurface into spacetime, it must carry a tensor that provides the second fundamental form, as discussed in Sect. 4.4.8. This must satisfy not only the three-dimensional analog of the Gauss equation (4.169), but also corresponding equations in the normal direction, known as the Codazzi-Mainardi equations (see, e.g., Eschenburg and Jost (2013)).

Yvonne Choquet-Bruhat[35] demonstrated that the initial value problem can be solved. That is, given the metric tensor and its derivatives on a Cauchy hypersurface, one can find a solution to (5.11) that evolves this data forward in time. Later, in collaboration with Robert Geroch, she showed in Choquet-Bruhat and Geroch (1969) that this solution exists uniquely for all future times.[36]

---

[34] For the general theory of partial differential equations, see Jost (2013).

[35] Fourès-Bruhat (1952), publishing at the time under the name Fourès-Bruhat.

[36] These results are significantly more challenging than the corresponding results for linear hyperbolic differential equations. The diffeomorphism invariance, which manifests itself, as discussed, in the second Bianchi identity (5.8), plays a crucial role here. For technical details, results in the presence of matter, and further developments, see, for example, Choquet-Bruhat (2009), Friedrich and Rendall (2000).

However, according to current physical understanding, the original initial conditions could be singular, representing the unfathomable Big Bang. At this point, we venture beyond the domain of General Relativity. During Einstein's time, it was still an open question whether the universe originated from such a singularity. We will not enter further into the development of cosmology here.

5.2.4 The Theory of Relativity and the History of Physics

Before we delve into other aspects of the history of the reception of Riemann's ideas, let us pause for a moment to survey the position of Einstein's theory in the history of physics. The General Theory of Relativity addresses perhaps the most fundamental problem in physics, the problem of motion. According to Aristotle, motion was purposeful, with the circular motions of celestial bodies and the straight-line motion of falling objects belonging to qualitatively different domains governed by different laws of motion. For Aristotle, motion was a process. However, the Aristotelian theory faced difficulties in explaining projectile and free-fall motions.

The scholastic philosophers of the late Middle Ages struggled with the question of how such a process could be sustained. In particular, the question of why acceleration occurs during free fall, rather than deceleration, could not be satisfactorily resolved within this framework. The analysis of these difficulties led to the development of the *impetus theory* by Nicole Oresme (1325–1382) and Jean Buridan (1301–1358), who conceived of a kind of causality carried along by the motion.[37] (In the physics of Galileo[38] and Einstein, motion, by contrast, is a state, so the problem that troubled scholasticism no longer arises.)

When Copernicus assigned Earth the status of a planet within the solar system, the conceptual separation between terrestrial and celestial motions was eliminated. Kepler subsequently treated the motions of celestial bodies not merely geometrically but also physically, proposing the Sun as the force center of the planetary system. Around the same time, Galileo analyzed free-fall and projectile motions, introducing the principle of inertia, which characterizes uniform, unaccelerated motion.

Newton, as previously discussed, developed a unified theory of physical motions, encompassing both the inertial motions described by Galileo and the circular—or more precisely elliptical, as clarified by Kepler—trajectories of planets around the Sun.[39] In this system, the

[37] The extensive studies by Duhem (1914–1917) were corrected in significant ways by Maier (1949, 1952, 1958, 1968, 1939, 1940). Building on these, see also Dijksterhuis (1953, 2006), English translation (Dijksterhuis 1961).

[38] In particular, Koyré (1966), especially p. 102, refutes the continuity of development from the medieval impetus to Galileo's impulse, as claimed by Duhem (1914–1917).

[39] Among the extensive literature, we mention Alexandre Koyré's compilation (Koyré 1955).

Sun's gravitational attraction, acting without an intervening medium, explains the deviation of planetary paths from straight lines. This force represents an external disturbance, exerted through an unexplained action at a distance, which compels motions to deviate from their natural paths in absolute space.[40]

Interestingly, the factor determining a body's inertia—its tendency to remain in its natural motion—was exactly proportional to the factor determining its susceptibility to gravitational forces from other bodies. This indicated a deeper relationship than what was described by Newtonian theory. As discussed, Einstein resolved this problem by linking gravitation to the structure of spacetime. This required the Riemannian conception of geometry, with point-dependent metric variations reflecting the influence of masses in space. Together with the integration of space and time into a four-dimensional continuum, systematically developed by Minkowski, this forms the basis of Einstein's General Theory of Relativity.

In this theory, the spacetime continuum also possesses a variable metric of the Riemannian type.

5.3 Lie and the Theory of Symmetry Groups

Sophus Lie took up the considerations of Helmholtz and Riemann to determine the geometries in which objects can move freely within the framework of his theory of transformation groups. On the one hand via precursors like Moritz Pasch (1843–1930) this led to the axiomatic foundation of geometry by David Hilbert, which opened up a research direction that dominated large parts of the mathematics of the twentieth century. On the other hand, this led to the modern theory of symmetry and invariance, which is fundamental in quantum mechanics, for instance. In the theory of principal bundles over Riemannian manifolds, Riemannian geometry then is combined with the theory of Lie groups. This then becomes the formal language of theoretical elementary particle physics.[41] For this, the theories of Weyl and Cartan are essential.[42] Cartan combines the group theoretical considerations of Lie with the geometric concepts of Riemann. Lie groups carry a certain Riemannian metric, which is determined by their structure and characterized by the fact that it is left invariant

---

[40] Georg Wilhelm Friedrich Hegel (1770–1831), in his *Encyclopedia of the Philosophical Sciences* (see Hegel (1830) or (1978); §§ 262–271 are relevant here), regarded the concept of a body moving without forces or interactions with other bodies as nonsensical. For Hegel, in the absence of other bodies, one could neither sensibly attribute motion nor even existence to a body. He critiqued Newton's theories while praising Kepler. In Hegel's view, this contradiction was resolved by considering the fundamental motion of a body not as the absurd rectilinear inertial motion but as Kepler's elliptical motion around a center of gravity, ultimately around the center of mass of the universe. Hegel's ideas have been variably assessed in light of relativity theory; see, for example, positive evaluations by Wandschneider (1982), Hösle (1988), and Halper (2008).

[41] See e.g. Jost (2009).

[42] Here, I sketch the considerations of Weyl and Cartan from the historical perspective. The systematic aspect will be taken up in Sect. 5.4.

by the group operations. The group operators are considered as geometric operations of the group on itself. Multiplication of all group elements by a fixed group element h thus yields a transformation of the group G. Each element g is therefore transformed into the element hg. Since such a transformation leaves the metric invariant, it is an isometry of the group considered as a Riemannian manifold. If we now let the element h generating such a transformation vary in a subgroup H of G, we obtain a whole family of such transformations. For a given group element g, we obtain an orbit Hg of new group elements, namely all elements of the form hg, where h is contained in the subgroup H. If we identify now all the elements of such an orbit with each other, i.e., consider them as equivalent to each other, we obtain a so-called quotient space G/H. Such a space is called a homogeneous space, and like the group G itself, it carries a natural Riemannian metric with respect to which the group G acts by isometries. These homogeneous metrics have been studied systematically by Cartan. A particularly important subclass of homogeneous spaces are the so-called symmetric spaces, which, as the name suggests, are characterized by a particularly high degree of symmetry. These spaces have been classified in the works of Killing and Cartan.[43] In addition to such a geometric characterization, they also admit a purely group-theoretical description. Consequently, their structure becomes particularly rich.[44] It has then been found that on the one hand these symmetric spaces constitute the most important class of examples of Riemannian manifolds (for instance, the spheres and the hyperbolic spaces are symmetric), and on the other hand they also include spaces that are central for Klein's conception of geometry. In this way Cartan could harmonize the approaches of Riemann and Klein which in the nineteenth century had still been seen as competing. In addition, Cartan also developed an alternative to the Ricci tensor calculus, that of the moving coordinate frames. This makes some aspects of tensor calculus geometrically more transparent and formally easier. Nowadays, mathematicians working in or utilizing Riemannian geometry usually employ an invariant calculus that combines the formalism of the covariant derivative, which has evolved from the parallel transport of Levi-Civita and Weyl, with the differential form calculus developed by Cartan because the coordinate independent meaning of the geometric expressions becomes most transpartent in that calculus. Most physicists, however, continue to favor Ricci's tensor calculus, developed in Sect. 4.4.2, as a convenient formalism. When using the tensor calculus, one need not account for the geometric meaning of the symbols employed, but can apply the formalism in an almost mechanical and automatic manner.

[43] See Helgason (1978).

[44] For details, Jost (2017).

5.4 Weyl and the Concept of the Connection on a Manifold

Hermann Weyl, as already explained above on page 99, introduced the concept of an affine connection.[45] This also leads to a natural relation between Riemannian geometry and the theory of Lie groups, but in an entirely different direction than the symmetric spaces studied by Cartan, which are Riemannian manifolds defined by Lie groups. According to Klein's conception, a geometry is characterized by its invariances, namely the group of those transformations that leave the geometric structure unchanged. On a Riemannian manifold, the geometric structure is the metric. Invariance transformations would here be those transformations that leave the distances between points unchanged. Thus, if P and Q are two points in a Riemannian manifold, then the distance between the two images gP and gQ under a transformation g has to be equal to the distance between the two original points. But the concept of a Riemannian manifold is so general that for a given such manifold M, except for the trivial transformation that leaves all points fixed, no such distance-preserving transformation g needs to exist. Thus, the notion of a Riemannian manifold does not fit into the Klein scheme. Now Riemann's notion of distance is obtained from an infinitesimal concept, the quadratic form that allows us to quantify the lengths of tangent vectors (direction elements) and angles between such vectors in a given point P. However, this is a notion that leads to a Euclidean measure on the space of direction elements, the tangent space, at the point P. This is where the invariance group acts, the group of Euclidean motions. In this view, the crucial aspect of Riemannian geometry now is that this infinitesimal action varies from point to point. The relationship between these actions according to Weyl is then achieved by a connection, that is, the possibility of setting infinitesimal structures in different points in relationship by transport along connecting curves (see also the mathematical discussion in Sect. 4.4.7 following Definition 4.3). This relationship, however, depends in general on the choice of connecting curve. This effect is infinitesimally measured by the Riemann curvature tensor. A Riemannian manifold can be regarded in this approach thus as a set of points, to each of which an infinitesimal Euclidean structure is assigned, which then can be compared with those of other points in a path dependent manner. The important point is not so much the infinitesimal Euclidean structure at the individual points, because this is abstract and for all points the same, but rather the concrete possibility of comparison of these structures as encoded in the manifold structure. So, although identical as such, these structures at the individual points can be related to each other in a variable manner. Here, an important abstraction step offered itself to Weyl. A Euclidean structure is an example of a Klein geometry. The same procedure can also be performed when based on another Klein geometry. Riemann himself had clarified the difference between a manifold as an object that contains only positional relationships, and a Riemannian manifold which carries an additional metric structure. Applying now the described Weyl method on a manifold, we initially

[45] We refer to the literature cited in a footnote on p. 62; also Scholz (1999).

have only a linear infinitesimal structure, the structure of a vector space in which vectors can be added and stretched or compressed, but where one cannot yet assign a length to them. The comparison of the infinitesimal linear structures at the various points of a manifold leads to the Weyl concept of an affine connection. So this is more general than the concept of a metric connection which is connected to the structure of a Riemannian manifold. There are also intermediate cases. Particularly important are the conformal structures. Here angles can be measured, but no lengths. The transition from one point to another leaves a scalar factor undetermined. Weyl interpreted this as a gauge freedom, so that at each point the length scale can be independently calibrated or gauged. This idea became extraordinarily fruitful for the development of geometry and physics, although the approach that Weyl himself had developed in this way for a unified field theory was not successful. Weyl wanted to combine Einstein's theory of gravity with Maxwell's electrodynamics and needed a gauge freedom in the transition between points. However, since the resulting gauge factor depends on the connecting path, this led to unacceptable physical consequences. But later, when the approach was modified so that the gauge factor was no longer a length factor, but a phase factor, and also more general invariance groups and gauge possibilities were included, this created the Yang-Mills theory which became the foundation of modern elementary particle physics. This will be explained in more detail in the next chapter. Ironically, Weyl's motivation and point of departure was the general theory of relativity, but the development he launched led to the modern quantum field theory, which so far has not succeeded in including the general theory of relativity in its program of the unification of the physical forces.

In abstract terms, the aim of modern physics is to derive a part of the phenomenal world, which might at first appear and seem to be very heterogeneous, from a few basic principles. In particular, a good physical theory should contain as few as possible free, contingent parameters, i.e., parameters that are not specified within the theory, even though it seems that every physical theory needs some not derivable, contingent constants. For example, the speed of light simply relates the units of space and time measurements, and therefore only expresses measurement conventions.[46] But the standard model of elementary particle physics contains several constants that cannot be so easily explained, like the fine structure constant. It is perceived as unsatisfactory by today's physicists due to the relatively large number of such indetermined parameters, despite its impressive predictive power. This seems to be a problem for Riemannian geometry to be a description of physical space, for a curvature that varies from point to point then poses the problem of how to determine it. If one only says that the structure of the space is determined by its curvature then nothing is explained physically. This was the starting point of Helmholtz. He derived, as explained above, much further structural constraints of space from a simple principle, that of the free mobility of bodies in space. Since this principle, as stated by Helmholtz, necessarily leads to a space of constant curvature, there is only a single parameter that is not theoretically derived, but only empirically determinable, namely the value of this constant curvature.

[46] On this point, see for instance Jost (2021).

However, it was the far reaching vision of Riemann that the explanation of the quantitative relations given by the metric must be sought not intrinsically, but in external forces acting on it.[47] Thus, in Riemann's vision, the curvature tensor and thus the structure of space have to be determined from physical principles, which then eliminated the problem of contingent parameters. Initially, this idea was not understood or not taken seriously, until later when it was confirmed by Einstein in a spectacular manner. One exception was the British mathematician W.K. Clifford (1845–1879), who translated Riemann's text (see Chap. 3) and wrote "this variation of the curvature of space is what really happens in the phenomenon which we call the motion of matter".[48]

5.5 Spaces as Tools for the Geometric Representation of Structures

There is one more essential difference between Riemann and Helmholtz, between the "hypotheses" and the "facts", that is central for understanding modern physics, even if it did not play a prominent role in the reception history. Helmholtz's goal was an ontological one, in the sense that he wanted to explore the nature and properties of the physical, the actual space in which we live and about which we obtain knowledge by gathering sensory data and performing physical measurements.[49] For Riemann, in contrast, a space is a mathematical structure, and the physical space is just one of many mathematically possible spaces. Therefore, Riemann's geometry can become the organizing principle of all possible "manifolds" of diverse, but comparable objects. Something like this had already indicated itself in the Cartesian coordinate descriptions and the phase spaces of mathematical physics as introduced and considered by Euler, Lagrange, Hamilton and Jacobi. Also, the introduction of the Gaussian complex plane can be seen in this light. This complex plane had inspired the idea of Riemann surfaces in Riemann's seminal work on complex function theory and on Abelian integrals. Relations between the (imaginary, not necessarily physically realized) objects or elements of any ensemble can then be represented and visualized by their position relative to each other in an abstract space. In the wake of Riemann, geometry could then penetrate almost all areas of mathematics, and this development continues in contemporary mathematics. The Hilbert space organizes the quantum mechanical states, Banach spaces contain the possible solutions of differential equations and variational problems, and Grothendieck conceived a geometric description of number theory, from which in recent decades major break-throughs in this mathematical field have emerged. The concept of the graph is used in various applications for the representation and visualization of possibly abstract relations between discrete elements.

---

[47] "We must seek the ground of its metric relations outside it, in binding forces which act upon it", Riemann, *Hypotheses*, Chap. III, p. 69.

[48] Clifford (1976), also in Clifford (1882, p. 21f).

[49] See, however, Schiemann (1997) for an analysis of the transition from an ontological to a phenomenological conception of physics also in the views of Helmholtz.

Modern theoretical high-energy physics interprets the results of scattering experiments of elementary particles as representations of invariance groups describing these particles in a vector space. In the contrasting approaches to the conceptual unification of the known physical forces, this phenomenological approach clashes with the still partly ontologically oriented theory of general relativity, without a definitive solution having emerged so far. It is probably an irony in the history of science that the Riemannian approach became fundamental for the perspective of the ontologically oriented and committed theory of Einstein that is aimed at uncovering the structure of space-time and that therefore can still be seen as partly ontologically oriented and committed,[50] while Lie's theory of invariance groups, which partially emerged from the desire of a mathematical clarification of the ontological approach of Helmholtz, found its way into the phenomenological perspective of quantum field theory whose spatial constructions are purely hypothetical in nature.

5.6 Riemann, Helmholtz and the Neo-Kantians

However, this already anticipates later developments which will be illustrated in more detail below, and we turn now to the initial reception of Riemann and Helmholtz.

As already mentioned, orthodox Kantians at first rejected the considerations of Riemann and Helmholtz. They were concerned with the three-dimensionality of space and its infinite extension as well as the role of non-Euclidean geometry. However, the rejection was not entirely unanimous. One, at the time very influential, group of spiritualist natural philosophers took up the idea of a four-dimensional space with great enthusiasm. When in England a popular magician by an apparently never fully uncovered trick made people believe that he could convert left to right-handed objects, it was thought that he achieved this by moving objects in an additional fourth spatial dimension,[51] that is, that he was a medium with access to the fourth dimension.[52]

Only with Einstein's general theory of relativity did Riemann's central idea, the question of the foundation and determination of the metric of space, come into the center of the

[50] We should recall, however, that through realizing the futility of the hole argument (see 133 above), Einstein, Schlick, Weyl, Eddington and others also moved away from an ontological conception of spacetime.

[51] See also above, on p. 24, the analysis of the Kantian argument of the relationship between handedness and spatial structure.

[52] The spiritualist medium was Henry Slade (1840–1904). Among the scientists who were taken in by him, was, for instance, Karl Friedrich Zöllner (1834–1882), the founder of astrophysics, who thereby ruined his scientific reputation. For details, we refer to Thiele (2003), Volkert (2018). Helmholtz, however, remained skeptical. A contemporary presentation can be found in Klein (1926, 1979), and for example a fairly free story by the theoretical physicist Kaku (1994) who presents the possibility of higher space dimensions as the essential and at the time sensational discovery of Riemann. A systematic mathematical analysis of spaces of arbitrary dimension had already been conducted before Riemann in a different context by Grassmann (1844), a work which founded linear algebra.

discussion. A later generation of philosophers tried on this basis to incorporate the arguments of Riemann and Helmholtz into the Kantian system.[53] Ernst Cassirer and Hans Reichenbach are prominent representatives of an attempt at a philosophical penetration of the theories of Riemann and Einstein.

We now turn to some of these intellectual directions.

5.7 The Axiomatic Foundation of Geometry

Sophus Lie presents and treats the problem of an axiomatic foundation of geometry under group theoretical aspects.[54] Since he seeks concepts that are basic and as elementary as possible, Riemann's approach appears less suitable for his purposes than Helmholtz'. Riemann obtains the local properties of space by integrating the infinitesimal line element, and neither line element nor integration are sufficiently elementary notions for axiomatic purposes. Helmholtz, while starting from elementary axioms about the mobility of bodies in space, is criticized by Lie because he proceeds in a mathematically unjustified fashion from the local to the infinitesimal properties of transformation groups and on top of that does not possess the appropriate group concept. In addition, the monodromy axiom as established by Helmholtz turns out to be superfluous, as already contained in the other axioms.

Lie presents his own set of axioms about the free infinitesimal mobility of bodies in space, and proves that a space which allows such flexible mobility in three and higher dimensions is necessarily either locally Euclidean, hyperbolic or spherical (in the terminology of his time, these two latter geometries were combined under the label non-Euclidean). Therefore, it is a space of constant Riemannian curvature, but this interpretation was not pursued by Lie. In two dimensions, however, there are other possibilities. In any case, the transition to more than three dimensions is already a mathematical matter of course for Lie that requires no longer any justification or discussion of physical or philosophical nature. If the infinitesimal assumptions are replaced by local ones, the problem becomes more difficult, and Lie succeeds only in solving it in the three-dimensional case.[55] It is the central assumption for Lie that the possible motions of a body constitute a group, meaning that the successive application of two motions yields again a motion and any motion can be reversed by applying its inverse.

---

[53] Martin (1969, 1972), Meinecke (1906), Natorp (1921), Nelson (1906).

[54] Lie ([S. 284–321]1890), Lie ([Dritter und Letzter Abschnitt, Abtheilung V]1893). Lie stated himself that he had been made aware of the work of Riemann and Helmholtz already in 1869 by Klein, pointing out that in these studies the concept of a continuous group was implicitly contained, but he himself did not turn to the considerations of Riemann and Helmholtz until 1884, when he had already worked out systematically his own theory of continuous groups (Lie ([Dritter und Letzter Abschnitt, S. 397]1893)). Somewhat strangely, in Hawkins (2012) Helmholtz does not appear in the presentation of the mathematical development of Lie, but only in that of Killing.

[55] Lie ([Dritter Abschnitt, S. 498–523]1893).

It would be incorrect, however, to see in Lie the mathematician that translated the ideas about the structure of space which Helmholtz had imprecisely formulated and formally unsatisfactorily elaborated into a mathematically exact form. Lie rather turns the problem around. Helmholtz wanted to derive the structure of space from empirically justified axioms. Lie, in contrast, from the outset wants to provide an axiomatic foundation of a particular class of geometries: "The *Riemann-Helmholtz* problem ... requires the identification of those properties of the family of Euclidean and the two families of non-Euclidean motions that are common to them and which distinguish these three families from all other such families."[56] For Lie, the aim of the axiomatics is no longer the metric of space, but a characterization of the motion group. This naturally fits into the context of Lie's research program, the theory of transformation and symmetry groups. The intentions of the theory of invariance groups of Felix Klein went in a similar direction. Here, however, I shall not analyze the complex relationship between the programs of Lie and Klein.

The axiomatic foundation of geometry was then developed most notably by David Hilbert.[57] Hilbert in his *Grundlagen der Geometrie* (Foundations of Geometry) lists five groups of axioms, which together found three-dimensional Euclidean geometry. These are the axioms of

1. *composition*, which connects the basic terms point, line and plane (e.g., that any two distinct points lie on exactly one line),
2. *arrangement*, which in particular defines the term *between* and stipulates that a line that enters into a triangle will also exits from that triangle,
3. *congruence*, which also defines the notion of motion and makes the comparison of distances and angles possible,
4. *parallels*, the axiom that is equivalent to the old Euclidean parallel postulate that through a point outside a line there is precisely one line that does not intersect the former,
5. *continuity*, firstly, the so-called *Archimedean axiom* that one, when sufficiently often repeating a given reference distance, can cover any other given distance, and second, the

[56] "Das *Riemann-Helmholtzsche* Problem ... verlangt die Angabe solcher Eigenschaften, die der Schaar der Euklidischen und den beiden Schaaren von Nichteuklidischen Bewegungen gemeinsam sind und durch die sich diese drei Schaaren vor allen anderen möglichen Schaaren von Bewegungen auszeichnen." (my translation) from Lie (1893, Dritter Abschnitt, S. 471) (emphasis in the original), and a similar formulation p. 397 ibd.

[57] Hilbert (1999), *Grundlagen der Geometrie,* Leipzig, Teubner, 1899; 13th ed. Stuttgart, Teubner, 1987 (with 5 supplements, in which several articles of Hilbert are reprinted, as well as supplements by Paul Bernays) and 14th ed. Leipzig, Teubner, 1999, with the essay Michael Toepell, *Zur Entstehung und Weiterentwicklung von David Hilberts Grundlagen der Geometrie,* that treats the developments prior to and after Hilbert's axiomatic approach to geometry; concerning the 7th ed., see also Schmidt (1970). Furthermore Hallett and Majer (2004), which not only reprints the original 1899 version, but also the other publications of Hilbert on the foundations of geometry. Hilbert's original text was edited for the present series with an extensive historical commentary by Volkert (2015).

completeness axiom that the given system of points, lines and planes cannot be extended by adding further elements without violation of at least one of the other axioms.

Thus, the completeness axiom specifies that there is a maximal set of elements that satisfy the axioms. That this may be required, is however by no means self-evident, but is consistently possible, as Hilbert explained, only upon acceptance of the Archimedean axiom. The main aim of Hilbert is then the proof of consistency and independence of the axioms. The consistency is achieved by constructing a model in which all the axioms are valid. In the present case, the model is of course just the three-dimensional Euclidean geometry. The independence is shown by replacing one of the axioms with another and then constructing another consistent model. For example, the models of non-Euclidean geometry prove independence of the axiom of parallels from the others. Hilbert then investigated systematically which of the above axioms are required to prove basic geometric theorems and which ones can be dispensed with for the individual results. For example, the Euclidean theory of proportions does not require the Archimedean axiom.

Above, the axiom of continuity was placed at the end. In Appendix IV of his *Grundlagen* Hilbert conversely places this axiom at the beginning of his considerations and then obtains a new systematic approach to Lie's theory of transformation groups, which does not require the infinitesimal constructions of Lie, which must assume differentiability conditions. Overall, the approach of Hilbert led mathematics into a different direction than what Riemann or Helmholtz had had in mind. For Hilbert, axioms are more or less arbitrary stipulations, instead of hypotheses that are in need of and amenable to an empirical test.[58] Hilbert's criterion is instead the internal consistency of a collection of axioms. The further development of Hilbert's program of a formalization of all of mathematics, therefore, is not part of our subject. It should however be noted that Hilbert's objective of a formalization and the corresponding role of axioms in mathematics and partly also in physics has been discussed very controversially and continues to be so discussed.

Hilbert inspired the approach of Nicolas Bourbaki, the pseudonym for a group of French mathematicians, that was particularly influential in the 50s and 60s of the twentieth century. Bourbaki developed and carried out a program of the systematic foundation and construction of all of mathematics from basic axioms. These axioms are selected solely because of their internal coherence and their theory-generating power. Against the formal approach of Bourbaki, there were always different voices that pointed to the intuitive foundations of mathematics or to its motivation by physical facts and discoveries and critically put this

[58] Stekeler-Weithofer (2008), in contrast, analyzes the relationship between the formal logical validity and the truth of geometrical statements based on real constructibility propositions with recourse to Kant's concept of a synthetic a priori validity. This quote must suffice here as an example for a very extensive and controversial discussion.

against a pure formalism for its own sake. Also in modern theoretical high energy physics, the axiomatic approach to quantum mechanics[59] and quantum field theory[60] could not really establish itself.

5.8 Conventionalism

The eminent mathematician Henri Poincaré (1854–1912) developed the so-called conventionalism as an alternative to both the apriorism of Kant and the empiricism of Helmholtz.[61] His intention was to analyze how the concept of space and its geometry arises from the mental effort of comparing and organizing sensory data. According to Poincaré, geometry, however, is not an empirical science because it is not revised by sensory experience and because it is exact, unlike all empirically obtained statements, which are approximate. The criterion for geometry is instead the simplicity of the description of sensory experiences. In principle, these could be described geometrically in various ways, but most of these descriptions are far too complicated and are therefore discarded. This consideration also plays an important role in the ideas of Einstein.

Conventionalism was further developed in the first half of the twentieth century, particularly by Hans Reichenbach.[62] However, I do not find the arguments fully convincing. One argument central to conventionalism was already discussed above in connection with Helmholtz's considerations. It says, for example, that we cannot decide whether rigid measuring rods exist, because to determine this, we would again need other tools assumed to be rigid, and so on. However, this seems irrelevant to me because, as long as no physical difference can be determined between a situation in which rigid measuring rods move in a space of constant curvature and one in which space and rods deform equally, such a distinction has no physical content but merely refers to a different representation of the same underlying reality. This fact is elaborated in the interpretation of gravity as a field, significantly shaped by Weyl.[63] Alternatively, if we consider the issue from a geometrical perspective, we can draw on Riemann's fundamental insight that one and the same manifold, i.e., one and the same geometrical situation, can be represented differently in various coordinate systems. If we represent Euclidean space in curvilinear coordinates, the Euclidean lines appear curved. However, this does not constitute a different geometry but merely a different coordinate representation of the same geometry.[64] One of Riemann's essential results was that invariants,

---

[59] von Neumann (1932, 2018).

[60] see Wightman (1976).

[61] Poincaré (1902). See also the detailed analysis by Torretti (2012).

[62] See the detailed account by Carrier (2009).

[63] Weyl (1993) and especially the extensively commented new edition Weyl (2025).

[64] A good example is found in Carrier (2009). The hollow Earth theory posits that the Earth is a hollow sphere with the celestial sphere inside. Geometrically, one can move from standard Euclidean geometry to such a hollow geometry by an inversion at the surface of the Earth. In this process, the

i.e., quantities independent of the chosen representation, can be derived from different representations of the same geometrical situation. In Riemann's work, these were the curvature quantities, but the principle is more general. These quantities then describe the underlying geometry, while the non-invariant aspects of the coordinates are merely tools for representation. For instance, we use maps in an atlas to depict the curved Earth's surface, even though this inevitably leads to distortions, as such a flat two-dimensional representation is convenient for many purposes. The conventionalist argument, therefore, either merely states that the same geometrical or physical reality can be represented differently, in which case the simplest and clearest representation is obviously the best, or it confuses an invariant reality with a variable representation.

Helmholtz aimed to determine the actual geometry, or more precisely, the best representation following the above clarification, through the observation of physical forces. For this reason, the heliocentric planetary system of Copernicus is preferable to that of Ptolemy or, more accurately, to that of Tycho Brahe, in which the other planets also orbit the Sun, but the Sun orbits the Earth. This is because the Sun, not the Earth, is the force center of the system.[65]

While a standard geometry like Euclidean or hyperbolic geometry has a specific coordinate choice that represents geometric realities most simply, this no longer necessarily holds in a more general geometric situation, such as in the context of general relativity. Reichenbach, therefore, proposed criteria for choosing the representation and experiments to test the infinitesimal deformation of physical objects to empirically address the question of the rigidity of rods and objects.[66]

5.9 Abstract Space Concepts

Modern mathematics further proceeded and advanced in the conceptualization of space from the basis established by Riemann.[67] Starting from the ideas of Riemann, Richard Dedekind and Georg Cantor (1845–1918) developed the concept of a set, a more abstract concept than that of a manifold.[68] A set is simply a collection of elements,[69] initially without further

---

infinitely distant point in Euclidean space is mapped to the center of the sphere. If the laws of motion in Newtonian mechanics are transformed accordingly (using tensor calculus), all physical laws of mechanics remain valid as before, and no empirical difference can be identified. The same physical reality is thus represented in other coordinates. Since this is a nonlinear coordinate transformation, the laws of motion become more complex in these new coordinates, making the Euclidean coordinates preferable. That is all. The question of whether hollow geometry is the "real" geometry is meaningless in this context because it confuses reality with its description. In any case, the theory was essentially abandoned after pictures of the earth from outer space were presented.

[65] This argument had already been advanced by Newton, see a footnote on p. 20.

[66] Hans Reichenbach (1928), reprinted Reichenbach (1979, Vol. 2).

[67] A systematic survey can be found in Jost (2015).

[68] On the history of the set concept, see for example Ferreirós (1999).

[69] The foundational issues connected with the set concept are not relevant for our purposes.

structure. From a set G, one can then construct a topological space by defining neighborhood relations between the elements. Such a structure is characterized by axioms. For this purpose, certain subsets of the set are distinguished as *open*. The conditions that must be satisfied are that both the empty set and the entire set G itself are open and that furthermore the intersection of finitely many and the union of arbitrarily many open sets are open again. These are the axioms of a topological space, a fundamental mathematical concept that was introduced and developed by Felix Hausdorff (1868- 1942) (Hausdorff (2002, 1914)). Otherwise, everything is arbitrary, quite in Hilbert's sense. In particular, no substantive interpretation of this formal structure is required. Even trivial extreme examples are not excluded. For example, the open sets of a topology can consist only of G itself and the empty set, or vice versa, all subsets of G could be open. These examples are important for an understanding of the scope of the concept. Also, the n-dimensional Euclidean space becomes a topological space, when we declare all distance balls, that is, all sets $U(p, r)$ of points that have a Euclidean distance less than a certain positive number r from a given point p, as open, and then further, all sets obtained from iterated finite intersections or countable unions of such distance balls, for any choices of p and r. A mapping f between topological spaces is then called continuous if the inverse image of every open set U, i.e. the set of points which are mapped by f to U, is again an open set.[70,71] In particular, the concept of continuity is thus a topological, not a purely set-theoretic concept.

Other conditions that can be imposed, but which go beyond continuity, require additional structure on the topological space G. Here the mathematics of the twentieth century offered many opportunities and examined many structures. Based on Riemann's considerations, the formal concept of a manifold has been made formally precise by David Hilbert, Hermann Weyl and others.[72] A manifold M of dimension n is a topological space that is characterized by the following property: Locally, by local coordinates, it can be bijectively related to the model space, the Euclidean space of dimension n, and the various such possibilities, that is, different choices of local coordinates, depend continuously on one another. This is now no more a simple concept, and the example which we have invoked already several times may illustrate this. We look at the earth's surface, represented by a spherical globe. Portions of this globe can be represented as maps in an atlas. The image of a map is two-dimensional Euclidean, and one can pass from one to another map through a transformation of their mutual overlap that is continuous in both directions.

---

[70] This includes and generalizes the well-known Weierstrass ε-δ-criterion of analysis, see below.

[71] Extensive material on these notions and their history can be found in the new edition Hausdorff (2002) of Felix Hausdorff's *Grundzüge der Mengenlehre* (1914) with detailed commentaries on the background in Walter Purkert, *Historische Einführung,* and a description of the evolution of the neighborhood axioms in Frank Herrlich et. al. *Zum Begriff des topologischen Raumes.* Sect. 3.2, *Fundamentaleigenschaften von Umgebungssystemen,* treats the relationship discussed in the text with the neighborhood axioms in $\mathbb{R}^n$ historically, on the basis of Hausdorff's own presentation in his course of the summer term 1912.

[72] For a detailed historical analysis see Scholz (1999).

A few more comments on the mathematical problem of the manifold concept[73]: For general topological spaces, it makes no sense to speak of a dimension. The concept of dimension arises only from the coordinate reference to a model space underlying the manifold concept. That the dimension of a manifold is uniqely determined is not evident, however. There is, in principle, the possibility that a manifold could be coordinatized locally by Euclidean spaces of different dimensions. As already explained above on p. 54, Luitzen E.J. Brouwer (1881–1966) (Brouwer (1911)) in 1911 succeeded in excluding such an ambiguity. Thus, each manifold possesses a unique dimension. Felix Hausdorff, the founder of topological set theory, pointed out that a further condition has to be included in the axioms. Any two different points of the manifold must possess disjoint coordinates neighborhoods, i.e. the coordinate descriptions must be fine enough to separate points from each other.

Also, an alternative, combinatorial approach to the manifold concept has been developed. Here the manifold, instead of being covered by coordinates neighborhoods, i.e. being locally described by n independent functions, is seamlessly assembled from topologically identical pieces, the so-called simplices, which can only touch at their faces, but otherwise are disjoint. (This may seem reminiscent of the ideas of Plato in his *Timaios* described in Sect. 2.1.) For example, a two- dimensional manifold like the already discussed sphere can be assembled from small curvilinear triangles. In higher dimensions, however, difficulties arise the investigation of which led to the development of the field of combinatorial topology.

Finally, we speak of a differentiable manifold, as explained in Sect. 4.4.1, if the transitions between different coordinate systems are always differentiable. The remarkable thing about this concept is that the differentiable structure is not seen from consideration of a single coordinate system, but only from the relations between two coordinate systems. Thus, the condition means that different coordinate descriptions have to be structurally compatible with each other. A manifold thus carries a differentiable structure when it admits a set of coordinate descriptions that are structurally compatible with each other and cover the entire manifold. The question under which conditions this is possible gives rise to the mathematical field of differential topology.

A structure which is at first completely different is that of a metric space. This concept was introduced by Fréchet (1906) and systematically developed by Hausdorff (1914). Again, one starts from a set G and assumes that one can define a distance function, which assigns to two points P and Q from G a distance $d(P, Q)$. This distance then has to satisfy the following axioms: The distance between two different points is always positive (only the distance of a point to itself is zero). The distance is symmetric, i.e. the distance from P to Q is the same as that from Q to P. For any three points P, Q, R the triangle inequality has to hold, i.e., $d(P, Q)$ is not greater than the sum of $d(P, R)$ and $d(R, Q)$. These axioms are again fulfilled for the Euclidean distance. Thus, the Euclidean space becomes a metric space in the sense of this definition. Every metric space is a topological space. To see this, as explained above in the Euclidean case, we can define the distance balls $U(p, r)$ and all other sets generated from them by taking finite intersections and arbitrary unions as open

[73] For details, we refer to Scholz (1999).

sets to satisfy the axioms. A mapping f between metric spaces is then continuous when the usual ϵ-δ–criterion of analysis is fulfilled, i.e. when for every ball of radius $\epsilon > 0$ in the image of f, we can find a ball of some radius $\delta > 0$ that contains the preimage of the former ball under f. In other words, we need to be able to always achieve that the images of two points under f have an arbitrarily small distance from each other, as long as these points themselves have a small enough distance.

Any Riemannian manifold is a metric space, since a metric in the sense of Riemann's conditions on a differentiable manifold generates a distance function that satisfies the above axioms. However, the local coordinate representations of a differentiable manifold are not given in metric terms, because the Euclidean distances in the local Euclidean charts need not coincide with those on the manifold itself. In our example above, this is the problem of cartography, namely that the mapping between the globe and the chart in the atlas does not and, in fact, cannot, preserve distances, but necessarily distorts some distance ratios.

Similarly, modern mathematics has axiomatically introduced a variety of different geometric structures. This approach, which is particularly associated with Hilbert, was systematised by Bourbaki (the pseudonym of a group of French mathematicians) after the 2nd World War, as already mentioned above, and declared the basis for all of mathematics. Although later, counter-movements were formed and the influence of this structural and axiomatic direction has by now significantly declined, it has nevertheless influenced the development of mathematics in many ways, especially in the areas of algebraic geometry, arithmetic and functional analysis. As outlined, Riemann, by his elaboration of abstract conceptual aspects, should be considered as the first pioneer of modern structure mathematics. In Riemannian geometry itself, however, this abstract approach has then become less important, at least in recent times. Here a significant question guiding research is the relationship between the curvature of a Riemannian manifold, which is an infinitesimal quantity, and the global topological structure of this manifold, that is, the relationship between the two main basic concepts that Riemann had introduced. This will be discussed in more detail in Sect. 6.1.

References

Abbott, E. A. 1884. Flatland: A romance of many dimensions. In *Nineteenth Century Science Fiction*, 57–68. Routledge (Reprinted, with an introduction by A, 1998. Lightman: New York etc., Penguin).

Beltrami, Eugenio. 1868. Saggio di interpretazione della geometria non-euclidea. *Giornale di Matematiche*, VI: 284–312.

Beltrami, Eugenio. 1868. Teoria fondamentale degli spazii di curvatura costante. *Annali di Matematica pura ed applicata* II: 232–255.

Beltrami, Eugenio. 1868. *Teoria fondamentale degli spazii di curvatura costante memoria*. F. Zanetti.

Breidbach, Olaf. 1997. *Die Materialisierung des Ichs: Zur Geschichte der Hirnforschung im 19. und 20. Jahrhundert*. Suhrkamp, Frankfurt/M.

Brouwer, Luitzen E. J. 1911. Beweis der Invarianz der Dimensionszahl. *Mathematische Annalen*, 11: 161–165.

Carrier, Martin. 1994. Geometric facts and geometric theory: Helmholtz and 20th-century philosophy of physical geometry. In *Universalgenie Helmholtz. Rückblick nach 100 Jahren*, ed. Krüger, Lorenz, 276–291. Berlin: Akademie-Verlag.

Carrier, Martin. 2009. *Raum-Zeit*. Walter de Gruyter.

Cassirer, Ernst. 1902. *Leibniz' System in seinen wissenschaftlichen Grundlagen*. Felix Meiner, 1998, based on the ed.

Choquet-Bruhat, Yvonne. 2009. *General Relativity and the Einstein Equations*. Oxford: Oxford University Press.

Choquet-Bruhat, Yvonne, and Robert Geroch. 1969. Global aspects of the Cauchy problem in general relativity. *Communications in Mathematical Physics* 14: 329–335.

Clifford, William Kingdon. 1882. *Mathematical Papers by William Kingdon Clifford:* Edited by Robert Tucker, With an Introduction by HJ Stephen Smith. Macmillan and Company.

Clifford, William Kingdon. 1976. On the space-theory of matter. In *The Concepts of Space and Time: Their Structure and Their Development*, 295–296. Berlin: Springer.

De Risi, Vincenzo. 2007. *Geometry and Monadology: Leibniz's Analysis Situs and Philosophy of Space*. Birkhäuser.

Dijksterhuis, Eduard Jan. 1961. *The Mechanization of the World Picture*. Oxford University Press.

Dijksterhuis, Eduard J. 1953. De Mechanisering van het Wereldbeeld. *Tijdschrift Voor Filosofie*, 15(1).

Dijksterhuis, Eduard J. 2006. *De Mechanisering van het Wereldbeeld*. Amsterdam University Press.

Duhem, Pierre. 1914–1917. *Le système du monde: histoire des doctrines cosmologiques de Platon à Copernic*, 5 Vols. A. Hermann.

Earman, John, and John Norton. 1987. What price spacetime substantivalism? The hole story. *The British Journal for the Philosophy of Science* 38 (4): 515–525.

Engler, Fynn Ole, and Jürgen Renn. 2018. *Gespaltene Vernunft: vom Ende eines Dialogs zwischen Wissenschaft und Philosophie*. Berlin: Matthes & Seitz.

Eschenburg, Jost-Hinrich, and Jürgen Jost. 2013. *Differentialgeometrie und Minimalflächen*, 3rd ed. Berlin: Springer.

Ferreirós, José. 1999. *Labyrinth of Thought: A History of Set Theory and Its Role in Modern Mathematics*. Basel: Birkhäuser.

Fourès-Bruhat, Yvonne. 1952. Théorème d'existence pour certains systèmes d'équations aux dérivées partielles non linéaires. *Acta Mathematica* 88: 141–225.

Fréchet, Maurice. 1906. Sur quelques points du calcul fonctionnel. *Rendiconti del Circolo Matematico di Palermo* 22: 1–74.

Friedrich, Helmut, and Alan Rendall. 2000. The Cauchy problem for the Einstein equations. In *Einstein's Field Equations and Their Physical Implications: Selected Essays in Honour of Jürgen Ehlers*, 127–223. B. Schmidt (Hrsg.). Berlin: Springer.

Gauss, Carl Friedrich, Bernhard Riemann, and Hermann Minkowski. 2013. *Gaußsche Flächentheorie, Riemannsche Räume und Minkowski-Welt*. Herausgegeben und mit einem Anhang versehen von J. Böhm und H. Reichardt. Teubner, Leipzig, 1984; Nachdruck Springer.

Gerhardt, Carl Immanuel, ed. 1855–1860. *Leibnizens mathematische Schriften*, vol. III–VII. H. W. Schmidt, Halle a. d. S.

Grassmann, Hermann. 1844. *Die lineale Ausdehnungslehre*. Leipzig: Otto Wigand.

Hallett, Michael, and Ulrich Majer. 2004. *David Hilbert's Lectures on the Foundations of Geometry, 1891–1902*, vol. 1. Berlin: Springer.

Halper, Edward C. 2008. Hegel's criticism of Newton. In *The Cambridge Companion to Hegel and Nineteenth-century Philosophy*. Cambridge University Press.

Hausdorff, Felix. 2002. *Felix Hausdorff – Gesammelte Werke Band II: Grundzüge der Mengenlehre*, ed. Brieskorn, Egbert, and Walter Purkert. Berlin: Springer.

Hausdorff, Felix. 1914. *Grundzüge der Mengenlehre*. Leipzig: von Veit.

Hawkins, Thomas. 2012. *Emergence of the Theory of Lie Groups: An Essay in the History of Mathematics 1869–1926*. Springer.

Hawkins, Thomas. 1984. The Erlanger Programm of Felix Klein: Reflections on its place in the history of mathematics. *Historia Mathematica* 11 (4): 442–470.

Hegel, Georg Friedrich Wilhelm. 1978. Enzyklopädie der philosophischen Wissenschaften, Bd.2. Suhrkamp, Frankfurt, M. mit den mündlcihen Zusätzen aus den Vorlesungen Hegels, hrsg von E. Moldenhauer und K. M. Michel.

Hegel, Georg Friedrich Wilhelm. 1830. *Enyzklopädie der philosophischen Wissenschaften*, 8th ed. Felix Meiner (hrsg. von F. Nicolin und O. Pöggeler auf der Grundlage der Fassung von 1830).

Helgason, Sigurdur. 1978. *Differential Geometry*. Lie Groups and Symmetric Spaces: Academic.

Helmholtz, Hermann. 1856–1857. *Handbuch der Physiologischen Optik*. Leopold Voss, Hamburg, Leipzig, 3 Bde.

Helmholtz, Hermann. 1863. *Die Lehre von den Tonempfindungen als physiologische Grundlage der Musik*. Braunschweig: F. Vieweg.

Hilbert, David. 1999. *Grundlagen der Geometrie*. Teubner, Leipzig, 1899; 13th ed. Stuttgart, Teubner, 1987 (with 5 appendices, reprinting various papers of Hilbert, and supplements of Paul Bernays) and 14th ed. Leipzig, Teubner, 1999, with the essay Michael Toepell, Zur Entstehung und Weiterentwicklung von David Hilberts Grundlagen der Geometrie.

Hösle, Vittorio. 1988. *Hegels System*. Hamburg: Felix Meiner.

Janssen, Michel, and Jürgen. Renn. 2022. *How Einstein Found His Field Equations: Sources and Interpretation*. Classic Texts in the Sciences. Berlin: Springer.

Jost, Jürgen . 2013. *Partial Differential Equations*, 3rd ed. Berlin: Springer.

Jost, Jürgen. 2009. *Geometry and Physics*. Berlin: Springer.

Jost, Jürgen. 2015. *Mathematical Concepts*. Berlin: Springer.

Jost, Jürgen. 2017. *Riemannian Geometry and Geometric Analysis*, 7th ed. Berlin: Springer.

Jost, Jürgen. 2021. Leibniz and modern physics. *Lo Sguardo: Rivista di Filosofia* 32 (1): 91–108.

Kaku, Michio. 1994. *Hyperspace: A Scientific Odyssey Through Parallel Universes, Time Warps, and the 10th Dimension*. Press: Oxford Univ.

Klein, Felix. 1872. *Vergleichende Betrachtungen über neuere geometrische Forschungen*. A. Düchert, Erlangen (reprinted Akad. Verlagsges., Leipzig, 1974).

Klein, Felix. 1921–1923. *Gesammelte mathematische Abhandlungen*, Hrsg. von R. Fricke und A. Ostrowski, 3 Bde. Julius Springer.

Klein, Felix. 1926. *Vorlesungen über die Entwicklung der Mathematik im 19. Jahrhundert*. Julius Springer.

Klein, Felix. 1928. *Vorlesungen über nicht-euklidische Geometrie*. Julius Springer.

Klein, Felix. 1972. Vergleichende Betrachtungen über neuere geometrische Forschungen. *Mathematische Annalen* 43(1): 63–100, 1893 (reprinted in K. Strubecker (Hrsg.), *Geometrie*, Darmstadt, Wiss. Buchges., 1972, pp. 118–155).

Klein, Felix. 1979. *Development of mathematics in the 19th century*, trans. from the German by M. Ackerman. Lie Groups: History, Frontiers and Applications, IX. Brookline: Mathematics Science Press.

Klein, Felix. 1871. Über die sogenannte Nicht-Euklidische Geometrie. *Mathematische Annalen* 4: 573–625.

Koenigsberger, Leo. 1902/03. *Hermann von Helmholtz*, 3 Bde. Braunschweig: Vieweg.

Koyré, Alexandre. 1955. *A Documentary History of the Problem of Fall from Kepler to Newton*. Philadelphia: University of Pennsylvania Press.

Koyré, Alexandre. 1966. *Études galiléennes*. Paris: Hermann.

Lie, Sophus. 1888–1893. *Theorie der Transformationsgruppen*. Unter Mitwirkung von F. Engel. Teubner, Leipzig.

Lie, Sophus. 1890. Über die Grundlagen der Geometrie. *Ber. Verh. königlich-sächsische Gesellschaft der Wissenschaften Leipzig, Math.-Phys. Classe* 42: 284–321.

Lorentz, H. A., A. Einstein, and H. Minkowski. 1982. *Das Relativitätsprinzip*, 8th ed. Wiss. Buchges., Darmstadt (reprint of 5th ed., 1923, B. G.Teubner).

Maier, Anneliese. 1949. *Die Vorläufer Galileis im 14. Jahrhundert. Studien zur Naturphilosophie der Spätscholastik*. Ed. di Storia e Letteratura, Roma.

Maier, Anneliese. 1952. *An der Grenze von Scholastik und Naturwissenschaft*, 2nd ed. Ed. di Storia e Letteratura, Roma.

Maier, Anneliese. 1958. *Zwischen Philosophie und Mechanik. Studien zur Naturphilosophie der Spätscholastik*. Ed. di Storia e Letteratura, Roma.

Maier, Anneliese. 1968. *Zwei Grundprobleme der scholastischen Naturphilosophie*, 3rd ed. Ed. di Storia e Letteratura, Roma.

Maier, Anneliese. 1939. *Das Problem der intensiven Grösse in der Scholastik*. Leipzig: Heinrich Keller.

Maier, Anneliese. 1940. *Die Impetustheorie der Scholastik*. Wien: A. Schroll.

Martin, Gottfried. 1969. *Immanuel Kant*, 4th ed. Berlin: de Gruyter.

Martin, Gottfried. 1972. *Arithmetik und Kombinatorik bei Kant*. Berlin: de Gruyter.

Meinecke, Wilhelm. 1906. Die Bedeutung der Nicht-Euklidischen Geometrie in ihrem Verhältnis zu Kants Theorie der mathematischen Erkenntnis. *Kantstudien* 11: 209–232.

Meulders, Michel. 2010. *Helmholtz: from enlightenment to neuroscience*, trans. from the French by L. Garey. MIT Press.

Minkowski, Hermann. 2013. *Space and Time: Minkowski's Papers on Relativity*, ed. Petkov, Vesselin. Montréal: Minkowski Institute Press.

Minkowski, Hermann. 1909. Raum und Zeit. Phys. Zeitschr. 10: 104–111.

Minkowski, Hermann. 1909. Raum und Zeit. *Jahresbericht der Deutschen Mathematiker-Vereinigung* 18: 75–88.

Natorp, Paul. 1921. *Die logischen Grundlagen der exakten Wissenschaften*, 2nd ed. Leipzig: Teubner.

Nelson, Leonard. 1906. Bemerkungen über die Nicht-Euklidische Geometrie und den Ursprung der mathematischen Gewißheit. *Abh. Friessche Schule, Neue Folge*, I: 373–430.

Noether, Emmy. 1918. Invariante Variationsprobleme. *Nachr. Königliche Gesellschaft der Wissenschaften Göttingen, Math.-phys. Klasse*, 2: 235–267.

Poincaré, Henri. 1902. *La science et l'hypothèse*. Paris: Flammarion (reprinted 1968).

Reichenbach, Hans. 1979. *Gesammelte Werke*, ed. Kamlah, Andreas. Braunschweig: Vieweg.

Reichenbach, Hans. 1928. *Philosophie der Raum-Zeit-Lehre*. Berlin und Leipzig: De Gruyter.

Rowe, David. 2025. *Felix Klein: Comparative Reflections on Recent Research in Geometry (Klein's "Erlangen Program")*. Birkhäuser: Classic Texts in the Sciences.

Russell, Bertrand. 1897. *An Essay on the Foundations of Geometry*. Cambridge University Press (Dover Publications, Reprint 1956).

Russell, Bertrand. 1899. Sur les axiomes de la géométrie. *Revue de Métaphysique et de Morale* 7 (6): 684–707.

Schiemann, Gregor. 1997. *Wahrheitsgewissheitsverlust. Hermann von Helmholtz' Mechanismus im Anbruch der Moderne. Eine Studie zum Übergang von klassischer zu moderner Naturphilosophie*. Darmstadt: Wiss. Buchgesellschaft.

Schmidt, Arnold. 1970. Zu Hilberts Grundlegung der Geometrie, 404–414 in Hilbert, David. *Gesammelte Abhandlungen*, 2nd ed., 1970, Bd.2. Berlin: Springer.

Scholz, Erhard. 1999. The concept of manifold, 1850–1950. *History of Topology*, 25–64.

Scholz, Erhard. 1999. Weyl and the theory of connections. In *The Symbolic Universe. Geometry and Physics 1890–1930*, ed. Gray, Jeremy, 260–284. Oxford University Press.

Sharpe. Richard W. 2000. *Differential Geometry: Cartan's Generalization of Klein's Erlangen Program*, vol. 166. Berlin: Springer.

Stachel, John. 1989. Einstein's search for general covariance. In *Einstein and the History of General Relativity*, volume 1 of *Einstein Studies*, ed. Howard, Don, and John Stachel, 63–100. Birkhäuser, 1980. Paper first read at the Ninth International Conference on General Relativity and Gravitation, Jena.

Stekeler-Weithofer, Pirmin. 2008. *Formen der Anschauung*. Walter de Gruyter.

Thiele, Rüdiger. 2003. Fechner und die Folgen außerhalb Naturwissenschaften. In *Interdisziplinäres Kolloquium zum 200. Geburtstag Gustav Theodor Fechners*, 67–111. Max Niemeyer, Tübingen.

Torretti, Roberto. 2012. *Philosophy of Geometry from Riemann to Poincaré*. Berlin: Springer.

Volkert, Klaus. 2015. *David Hilbert: Grundlagen der Geometrie (Festschrift 1899)*. Klassische Texte der Wissenschaft. Springer.

Volkert, Klaus. 2018. *In höheren Räumen: der Weg der Geometrie in die vierte Dimension*. Berlin: Springer.

von Helmholtz, Hermann. 1883. *Wissenschaftliche Abhandlungen: Physikalische Optik. Physiologische Optik. Physiologische Akustik. Erkenntnisstheorie-Physiologie*, vol. 2. Leipzig: Ambrosius Barth.

von Helmholtz, Hermann. 1910. *Handbuch der Physiologischen Optik*, vol. III. Hamburg: Leopold Voss.

von Helmholtz, Hermann. 1921. *Schriften zur Erkenntnistheorie*. With comments by Moritz Schlick and Paul Hertz, ed. Ecke Bonk. Springer (Reprint Springer, Wien/New York, 1998).

von Neumann, John. 1932. *Mathematische Grundlagen der Quantenmechanik*. Berlin: Springer.

von Neumann, John. 2018. *Mathematical Foundations of Quantum Mechanics: New Edition*. Princeton University Press.

von Waltershausen, Sartorius. 1856. *Gauß zum Gedächtnis*. Leipzig: S. Hirzel.

Wandschneider, Dieter. 1982. *Raum, Zeit, Relativität*. Frankfurt: Klostermann.

Weyl, Hermann. 1952. *Space, Time, Matter*. Dover (translated from the German).

Weyl, Hermann. 1993. *Raum, Zeit, Materie: Vorlesungen ber allgemeine Relativitätstheorie*, 8th ed. ed. Ehlers, Jürgen. Berlin: Springer.

Weyl, Hermann. 2025. *Raum, Zeit, Materie: Vorlesungen ber allgemeine Relativitätstheorie*. Klassische Texte der Wissenschaft. New ed. with commentary by Domenico Giulini and Erhard Scholz. Berlin: Springer.

Wightman, Arthur S. 1976. Hilbert's sixth problem: mathematical treatment of the axioms of physics. In *Proceedings of Symposia in Pure Mathematics*, vol. 28, 147–240. AMS.

6 Modern Research

Riemannian geometry today is a central and essential part of mathematics, with many connections to other fields. This is uncontroversial. The philosophical debates and controversies are largely decided. Although modern physics is still struggling with the fundamental problem of the unification of all forces, specifically on one hand the electromagnetic, weak and strong interactions, which are already unified in the so-called Standard Model, and gravity on the other hand, it is also undisputed that Riemannian geometry provides an essential formalism for this purpose.

A sketch of the state of research can therefore only mean to outline the basic ideas and statements of the various contemporary research directions to the extent that this is at all possible without the use of a specialized formalism and a correspondingly developed terminology.

The purpose of this section can thus consist only in explaining the key concepts and results, but not in tracing their historical development. For details and literature references, we need to refer to the monographs and surveys listed in the selected bibliography.

6.1 The Global Structure of (Riemannian) Manifolds

A central question in recent research is the relationship between the topological structure of a manifold and the Riemannian metrics it can support. As we discussed earlier when presenting Riemann's considerations, the surface of the sphere—a specific two-dimensional manifold—cannot support a metric of negative curvature. It is natural, then, to pose similar questions in higher dimensions. However, it is necessary to clarify what is meant by negative or positive curvature in this context, as Riemannian curvature in higher dimensions is described by a tensor rather than a single number. From this tensor, various scalar values can be derived in different ways.

© The Author(s), under exclusive license to Springer Nature Switzerland AG 2025

J. Jost, *Bernhard Riemann — On the Hypotheses Which Lie at the Bases of Geometry*, Classic Texts in the Sciences, https://doi.org/10.1007/978-3-031-91489-8_6

The most important approach, which also corresponds to how Riemann conceptualized the curvature tensor, consists in determining the curvature of two-dimensional substructures of the manifold. This refers to the so-called sectional curvature, i.e., the curvature of infinitesimal planes spanned by two independent directions, as explained in Sect. 4.4.9. Since these planes are surfaces, i.e., two-dimensional entities, their curvature reduces to a single real number. One then says that the Riemannian manifold has negative sectional curvature if the curvature is negative for all such planes at all points. With these concepts, one can show, for example, that the higher-dimensional analogs of the spherical surface, the so-called spheres, also cannot carry a metric of negative curvature. In general, the existence of a metric with sectional curvature of fixed sign, whether negative or positive, leads to strong topological constraints on the underlying manifold. This is important for understanding possible space structures.

The theory of negatively curved metrics also has a close connection to the theory of dynamical systems. The reason lies in the fact that in negative curvature, geodesic curves, i.e., shortest connections, analogous to Euclidean straight lines, which begin at the same point, diverge exponentially rather than only linearly as in the Euclidean case. This exponential divergence corresponds precisely to the exponential magnification of even the smallest differences, which is characteristic of so-called chaotic dynamics. The geodesic flow, i.e., the tracing of geodesic lines, in spaces of negative curvature is thus an example of a chaotic dynamical system. Consequently, the mathematical methods developed for this purpose can be applied in geometry, and conversely, geometry provides an important example of a chaotic dynamical system through which new insights into chaos can be gained.

The theory of Riemannian manifolds with positive curvature, on the other hand, leads in a completely different direction. If the curvature is not only positive but also nearly constant, then the underlying space must already have the topological structure of a sphere, as has been known since the fundamental sphere theorems of Wilhelm Klingenberg (1924–2010) and Marcel Berger (1927–2016) from the 1960s. Nevertheless, the theory of spaces with positive curvature today appears far less complete than that of spaces with negative curvature. Spheres themselves, however, even carry a metric of constant curvature, and spaces of constant curvature represent important model spaces in geometry, with whose properties then those of other Riemannian manifolds can be compared. The classification of spaces of constant curvature, whether positive, negative, or zero—the so-called space forms—has long been completed. As Riemann and Helmholtz already recognized, these are precisely the spaces where the free mobility of rigid bodies is possible. However, the question here was essentially a topological or group-theoretical one. The problem arose from the fact that one can derive new spaces of constant curvature with more intricate topological shapes from the model spaces of the sphere, the Euclidean space, or the hyperbolic space through quotient formation. Let us consider two-dimensional examples for illustration, which, however, can be carried out in the same way in any dimension.

We take the two-dimensional sphere and identify diametrically opposite points, the so-called antipodes, with each other. Thus, we obtain a new space, the so-called projective plane or the space of elliptic geometry, in which each point corresponds to a pair of points, namely an antipodal pair, on the sphere. The group-theoretical aspect of this construction arises from the fact that the motion of the sphere, which maps every point to its antipode, preserves the distance relations, as the distance between two points is the same as the distance between their antipodes. Such a motion of a space that keeps the distance relations invariant is called an isometry. The isometries form a group because, in particular, the successive execution of two isometries is again an isometry. This was the foundational idea behind the theories of Felix Klein and Sophus Lie.

Similarly, the translations of the Euclidean plane form a group. A subgroup of this translation group, for example, is formed by those translations that change the two coordinates of a point by integer amounts (instead of arbitrary real values), as the successive execution of two such integer translations results in another of the same type. If one now identifies points in the plane that can be transformed into each other by such an integer translation, or equivalently, points whose coordinates differ only by integers, then one obtains a new surface with the connectivity properties of an annular ring. Such a surface is called a torus. Just as the Euclidean plane, a torus also carries a metric of zero curvature. Similarly, the hyperbolic, non-Euclidean plane also allows such quotients which then also carry metrics of negative curvature.

In particular, there is a close connection here with one of the most important mathematical theories developed by Riemann, the so-called Riemann surfaces. Each such quotient naturally carries the structure of a Riemann surface, and the totality of these surfaces leads to Riemann's concept of the moduli space. In any case, the classification of spaces of constant curvature, or in group-theoretical terms, the classification of discrete subgroups of the isometry groups of the sphere, Euclidean space, and hyperbolic space, has been solved by mathematicians.[1] The connection between Riemannian geometry and group theory, however, is even more general. In addition to the model spaces of constant curvature, there are other Riemannian manifolds with transitive isometry groups, i.e., those in which isometries can map any point to any other point. This leads to the classification theory of Lie groups, as isometry groups are transformation groups in the sense of Lie, and to the theory of symmetric spaces, the most important class of such spaces, and their quotients by discrete groups of isometries.

These theories were developed particularly by Wilhelm Killing, Élie Cartan, and Hermann Weyl. The symmetric spaces naturally form an important class of model spaces in Riemannian geometry. Moreover, they exhibit deep and fundamental connections to number theory, which were pivotal for mathematical research in the twentieth century. However, this will not be further elaborated here. The mention of the fact that we constructed the torus above using integers must suffice as a hint. In any case, this points to an essential, and probably the most important, unity of algebraic, geometric, and analytical structures that motivate modern mathematical research, decisively inspired by Riemann's work.

---

[1] See, for example, the collection (Freudenthal 1978), which, however, also delves into research areas somewhat outside the main directions of modern geometry.

We presented the Riemannian concept of sectional curvature as a way to describe the curvature behavior of a manifold using numbers. For a Riemannian manifold of dimension n, this method yields $n(n-1)/2$ numbers for each point of the manifold, as this is the number of independent plane directions at a point. By averaging, this can be reduced to fewer numbers. Averaging over all planes containing a common direction results in the so-called Ricci tensor, introduced in Sect. 5.2.3, which can then be described using n numbers, corresponding to the number of independent directions at each point. If one further averages over all directions, a single number at each point is obtained: the so-called scalar curvature. Integrating the scalar curvature over the points of the manifold yields a single number for the entire manifold, the so-called total curvature.

From the total curvature, normalized by the volume of the manifold, one obtains the so-called Einstein-Hilbert functional, which, as Hilbert demonstrated, not only leads to Einstein's field equations of general relativity but also reveals essential structures of Riemannian geometry.[2]

Naturally, each step of such averaging involves a coarsening of the description. Accordingly, the classes of objects become broader. For instance, there are significantly more manifolds that admit a metric with positive Ricci curvature or scalar curvature than those with positive sectional curvature. In dimensions greater than 2, as shown surprisingly by Joachim Lohkamp, every manifold admits a metric with negative Ricci curvature, so the existence of such a metric imposes no structural constraints. The situation is different for positive Ricci curvature. Somewhat surprisingly, the picture here is much clearer than for spaces with a Riemannian metric of positive sectional curvature, although the latter is a stronger condition than the former. But one does not understand well which further restriction that stronger condition entails. If we were to summarize the current state of research in Riemannian geometry and its extensions, then we should say that the structural theory of spaces that admit either negative (or, somewhat more generally, nonpositive) sectional curvature or positive Ricci curvature is quite well developed.

The study of Ricci curvature for manifolds of dimension 3 recently led to the solution of one of the most challenging problems in topology and one of the most famous problems in mathematics: the Poincaré conjecture, resolved by Grigori Perel'man. While not every three-dimensional manifold admits a metric with positive Ricci curvature, by modifying the metric towards constant Ricci curvature, the underlying manifold can be decomposed into parts that admit constant Ricci curvature, which can then be classified in three dimensions.

This reveals a fundamental idea, particularly developed by Shing-Tung Yau, which connects topology, geometry, and analysis, and has also led to the solution of numerous other important problems. The concept of a manifold itself does not include a metric. One might reframe this by noting that the same manifold, as a topological object, can carry many different Riemannian metrics. The fruitful idea is to select, among these many possible metrics, a particularly favorable one using an optimization principle. Finding such a metric often constitutes the main technical difficulty. However, such a metric, as the solution to

[2] See, for example, the discussion in Jost (2017, 2026).

an optimization problem, possesses certain properties that can then provide insights into the structure of the underlying manifold. It should be noted that this does not constitute circular reasoning, as proving the existence of an optimal metric already requires utilizing properties of the manifold. The optimal metric then allows deriving additional properties based on these initial ones, which would typically be much harder or impossible to obtain using alternative methods. Conversely, one can also use topological methods to derive many geometric insights about Riemannian manifolds, as demonstrated particularly by Mikhail Gromov.

6.2 Riemannian Geometry and Modern Physics

The concepts of Riemannian geometry are fundamental not only for the general theory of relativity, but also essential for modern quantum field theory and theoretical elementary particle physics, from the so-called standard model to the latest developments, such as (super)string theory.

To discuss this, we need an important generalization of the concept of a manifold, that of a fiber bundle. We had discussed a special case of this concept, the principal bundle, already above from a historical perspective in Sect. 5.3. Here we want to introduce it under a different, systematic point of view. As explained earlier, a manifold is a collection of distinct points with qualitative positional relationships. This concept can now be extended by taking another object instead of a point. Examples of such objects that are particularly important for geometry and theoretical physics are Lie groups and vector spaces. Such an object represents the model of what is called a fiber and a *fiber bundle* is then a collection of copies of the fiber, in a manner analogous to a manifold. If we suppress the structure of the fiber and understand those copies only as points, we obtain a manifold again. This manifold parametrizes the collection of fiber specimens. However, the relative positions of the various fibers have yet to be specified. This means that one must also specify how to pass from a specific element of one fiber to a certain element of another fiber. The concept that expresses this is called the connection of the fiber bundle. It can be seen as a generalization of the parallel transport of Levi-Civita, and it emerges from the subsequent ideas of Hermann Weyl. That parallelism expresses how to transport a directional element at a point into a directional element at another point, by parallel transport of this element along a given curve (see Definition 4.3 and the subsequent discussion in Sect. 4.4.7). The directional elements provide an important example of a fiber bundle, the so-called tangent bundle of the manifold. The directional elements at a point of the manifold form the fiber belonging to this point, called the tangent space of the point. The abstract fiber here is a vector space of the same dimension as the underlying manifold, as the number of linearly independent directions at a point, i.e. the dimension of the tangent space, provides just the dimension of the manifold. More generally, a fiber bundle whose fibers are vector spaces is a vector bundle.

We thus have seen one of the most important examples of a fiber, namely a vector space. The other example is a Lie group. The two examples are interdependent, because the structure-preserving transformations of a vector space form a Lie group, and conversely, a Lie group can operate on a vector space. One speaks here of a representation of the Lie group.

The Lie group as an abstract object thus becomes concrete through its operation, its representation on a vector space. This is now fundamental for theoretical particle physics. An elementary particle, or better, a particle type as the electron or a particular quark, is conceptualized by its symmetries and thus distinguished from other particles with other symmetries. The symmetries can in turn by expressed by a Lie group. But the particle is only realized through the action of this group on a vector space, and the observational data of particle scattering experiments are interpreted in this framework. The particle as such is invariant, but in the observation that invariance is broken, and one finds a particular element of the fiber of a vector bundle. The fiber thus expresses the various possible manifestations of the particle. This might now suggest a certain analogy with the concept of a manifold as a juxtaposition of points, even if the viability of this analogy is rather unclear and leads into basic issues of the unification of the fundamental forces. The elements of the fiber correspond to the different concrete manifestations of a particle which by its intrinsic nature is symmetric, that is, to the observed or possible breakings of this symmetry. In the same manner, a point in a Lorentzian manifold could likewise be interpreted as a concrete phenomenal appearance of a state that by itself is indifferent regarding its position in space and time.

However, the unification of the fundamental forces seems to be more difficult. One of the currently most popular approaches, string theory, no longer operates with pointlike particles, but its elementary objects, the strings, have the structure of a loop. Different particles then correspond to the different excitation or vibrational states of such strings. If such a loop, i.e. a one-dimensional object, moves in space-time, it sweeps out a surface which can be interpreted again as a Riemann surface. According to the principles of quantum mechanics, we cannot specify which surface is traversed, but only know that smaller surfaces, more precisely those surfaces with a smaller action integral, are more likely than larger ones. Therefore, we have to form a so-called Feynman integral over all possible Riemann surfaces. The underlying mathematical structure leads to a fascinating convergence of a wide range of mathematical fields. The consideration of an additional symmetry, the so-called supersymmetry, between bosonic or interaction particles on the one hand and fermionic or matter particles or on the other hand leads to superstring theory. This requires, however, for reasons of mathematical consistency, no longer a four-dimensional, but rather a ten-dimensional space-time continuum.[3] The six extra dimensions are thought to be so tiny that

[3] Ordinary string theory requires even 26 dimensions.

they are macroscopically not visible. Because of the occurring particle symmetries, these small spaces have to carry a certain Riemannian metric with vanishing Ricci curvature, named after their discoverers the Calabi-Yau metric.

Riemann had the vision of uniting geometry, physics, and natural philosophy. During his short life, he himself could not realize this dream. Today, 150 years later, we still have not fully achieved this unification, but we may have come somewhat closer to it. Whatever has been accomplished depends in essential ways on the fundamental concepts and is permeated by the remarkable ideas that were put forward by Riemann.

6.3 Metric Geometry and Geometric Methods in Machine Learning

The concept of a Riemannian manifold is also of great utility in machine learning. On the one hand, one can interpolate locally within a Riemannian manifold, and on the other hand, such a manifold can represent complex global relationships. The ability to interpolate locally is based on the fact that a Riemannian manifold resembles Euclidean space on a small scale, i.e., for sufficiently small distances. If one knows several points close to each other in a Riemannian manifold, one can conclude that the region between them behaves similarly to a region spanned by points with similar distance relationships in Euclidean space. However, on a large scale, a Riemannian manifold can have a rich topological structure, very different from Euclidean space. The combination of these properties makes the concept so useful. The Riemannian curvature mediates between small-scale behavior and large-scale structure.

In machine learning, however, one typically deals not with continua but with discrete datasets. To nevertheless make use of Riemannian geometry, one attempts to identify or construct a Riemannian manifold that can accommodate these points while at least approximately preserving their distance relationships. For this purpose, it is helpful if curvature measures can already be evaluated for discrete datasets. This will now be explained.

As outlined in Chap. 4, particularly in Sect. 4.4.7, curvature is the fundamental concept of Riemannian geometry. As Riemann demonstrated, a Riemannian metric is determined by its sectional curvatures, and these curvatures dictate the geometry of geodesic triangles, configurations where three points on a manifold are connected by shortest geodesics. For simplicity, we will refer to such structures simply as triangles.

This connection between sectional curvature and the geometry of triangles was central to Riemann's argument, though he focused on infinitesimal triangles. However, infinitesimal triangles approximate local triangles, so sectional curvatures, being infinitesimal quantities, determine the shapes of local triangles. Conversely, the shapes of local triangles can provide bounds for the sectional curvature of a Riemannian manifold. Since such triangles can also occur in more general metric spaces, this opens up the possibility of defining curvature—or at least curvature bounds—for such metric spaces, thereby generalizing the concepts of Riemannian geometry. Such ideas were developed by Karl Menger (1902–1985) (Menger 1929), Abraham Wald (1902–1950) (Wald 1936), Alexander Danilovic Alexandrov (1912–

1999) (Alexandrov 1957), and Herbert Busemann (1905–1994) (Busemann 1955), with Alexandrov's approach proving the most fruitful.[4]

To describe such generalized curvature concepts, we recall the notion of a metric space introduced in Sect. 5.9. For a triangle with vertices x_1, x_2, x_3 in a metric space (M, d) that is geodesic, that is, in which any two points can be connected by a shortest geodesic, we consider a Euclidean comparison triangle. This consists of three points $\bar{x}_1, \bar{x}_2, \bar{x}_3$ in the Euclidean plane, with the same pairwise distances as the points in M, i.e., $\|\bar{x}_i - \bar{x}_j\| = d(x_i, x_j)$ for all i, j. Then, for example, M is assigned nonpositive (nonnegative) curvature if the distance from x_i to a point on the geodesic connecting the other two points is not greater (less) than the corresponding distance in the Euclidean comparison triangle. By choosing a comparison triangle in a 2-dimensional Riemannian manifold with a different constant curvature—such as a suitably scaled hyperbolic plane or sphere, different curvature comparisons can be defined. In this sense, spaces of nonpositive curvature share similar properties with Riemannian manifolds of nonpositive curvature, leading to significant implications explored by Gromov (1980, 1999). For example, one can associate a metric space with a discrete group, the so-called Cayley graph. To do this, one selects a system of generators, i.e., elements from which any group element can be generated through multiplications, and then connects two group elements with an edge if one of them can be obtained from the other by multiplication with an element from this generating system. Gromov observed that most groups, or more precisely their Cayley graphs, have non-positive curvature in the Alexandrov sense. This has numerous consequences for the structural theory of discrete groups.

For geometric methods in machine learning, however, this framework is not general enough. While many methods construct Riemannian manifolds, they typically begin with a discrete set of data points. To apply geometric methods, this dataset must be metric, meaning there must a distance $d(p, q)$ to every pair of data points p, q be given or be constructed. Such metric datasets often arise when various properties of underlying objects are quantified. A data point then becomes a vector of these property values, viewed as a vector in a Euclidean space whose axes represent the values of the respective properties. The Euclidean distance between such vectors is then used.

If a large number of properties are evaluated, we work in a space of very high dimension. However, since property values are typically correlated, the data points do not spread throughout the entire space but rather lie on a lower-dimensional substructure. It is natural to assume that this structure is a manifold, hence in particular free from singularities. Singularities would complicate interpolation between data points. In contrast, a manifold locally resembles a Euclidean subspace, enabling straightforward interpolation. A Euclidean subspace of dimension n is uniquely determined by $n + 1$ points, meaning one knows exactly what lies between these points. For a manifold, this holds locally, in regions that are sufficiently small relative to the curvature. However, this does not hold globally, as an n-dimensional submanifold of an N-dimensional space can extend in far more than n direc-

[4] See, e.g., Aleksandrov et al. (1986), Burago et al. (2001), Jost (1997) for systematic treatments of such curvature bounds.

tions. Even a curve in space can wind in multiple directions. Thus, a manifold is a favorable structure for modeling datasets that are locally regular but potentially complex on a global scale.

We thus assume the data points lie on or near an n-dimensional submanifold M of an ambient $\mathbb{R}^N$, where N may be much larger than n. A submanifold of Euclidean space automatically inherits a Riemannian structure (see Sect. 4.4.8). The first task of geometric machine learning is to reconstruct this Riemannian manifold from the data points and their distances. (Strictly speaking, the assumption of an embedding in Euclidean space is unnecessary; a metric dataset suffices.)

How can this be achieved? A graph approximating the desired manifold is constructed by connecting sufficiently close data points by edges.[5] The more data points we have, the closer the points to be connected should be, and as more data points are added, the graphs should converge to the desired manifold. However, it is not immediately clear what it means for a sequence of graphs to converge to a Riemannian manifold. In Belkin and Niyogi (2003), an indirect approach is proposed: construct a Laplace operator Δ for the graph, which operates on functions defined on the data points. For a given function f and a data point p, Δ compares the value $f(p)$ with the average of the values at all points q that are connected to p by an edge. If we use $p \sim q$ to denote that p and q are connected, and $\deg p$ to represent the number of points connected to p, then we have the following expression:

$$(\Delta f)(p) = f(p) - \frac{1}{\deg p} \sum_{q \sim p} f(q). \tag{6.1}$$

Thus, Δ is a linear operator on a space of functions, and as such, it has eigenvalues and eigenfunctions. The eigenvalues and eigenfunctions of Δ for a sequence of graphs can then be analyzed to determine the limit space M.[6]

The operator Δ also induces a random walk on the underlying graph. A random walker located at point p at time $t \in \mathbb{N}$ selects one of the points connected to p to move to at time $t + 1$. Each of these connected points is chosen with equal probability $\frac{1}{\deg p}$. For a sequence of graphs constructed as described above, consisting of an increasing number of data points, these random walks converge to a stochastic process that characterizes the desired limiting manifold, as utilized in Hinton and Roweis (2002).

With such methods, a Riemannian manifold representing the dataset can be obtained. Even if its dimension n is small relative to the original dimension N, it might still be too large to visualize. To visualize it, one would like to represent this manifold in two dimensions (2D). Of course, this is generally not possible, so the goal is to create a visualization that preserves the essential qualitative properties. For a metric graph Γ, one can attempt to construct a graph Γ_0 in the plane that has the same set of nodes, with node distances as close

[5] A graph consists of a set of vertices, which in this case is the set of data points, and a set of edges. An edge is a pair of connected vertices.

[6] For a precise mathematical treatment, see, e.g., Joharinad and Jost (2023).

as possible to those in Γ. There are various methods to achieve this. However, it turns out that better results can be obtained by introducing an intermediate step.

For each data point p, a small graph, known as a *star*, is first constructed by connecting p to the k nearest data points $p_1, \ldots, p_k$ and assigning the corresponding edge lengths $d(p, p_j)$. This approach is reminiscent of the Riemannian normal coordinates discussed in Sect. 4.4.5. This metric can then be manipulated appropriately to better capture the local structure. However, these locally modified stars must then be reconnected into a global graph.

A corresponding method, UMAP from McInnes et al. (2018), utilizes abstract concepts from category theory to achieve this. The resulting visualization has proven to be very effective and is currently widely used, especially in cell biology.[7]

But so far, there is no curvature. This brings us to the next step: a concept of curvature for general metric spaces, as developed in Joharinad and Jost (2019, 2023). The geometric idea behind this is as follows. A characteristic property of Riemannian sectional curvature is that the smaller the curvature, the more easily intersection points of distance balls can be obtained.

For a point x in a metric space (X, d) and a radius $r \geq 0$, we consider the distance ball

$$B(x, r) = \{y \in X : d(x, y) \leq r\}.$$

We now examine the balls $B(x_i, r_i)$ for three points $x_1, x_2, x_3 \in X$, where the radii are defined as:

$$\begin{aligned}
r_1 &= \frac{1}{2}(d(x_1, x_2) + d(x_1, x_3) - d(x_2, x_3)) \\
r_2 &= \frac{1}{2}(d(x_1, x_2) + d(x_2, x_3) - d(x_1, x_3)) \\
r_3 &= \frac{1}{2}(d(x_1, x_3) + d(x_2, x_3) - d(x_1, x_2)).
\end{aligned}$$

These radii are chosen to be the smallest ones such that

$$B(x_i, r_i) \cap B(x_j, r_j) \neq \emptyset \tag{6.2}$$

is possible for all pairs (x_i, x_j). (In a complete Riemannian manifold, this is always the case.) However, in general, there is no triple intersection, meaning:

$$B(x_1, r_1) \cap B(x_2, r_2) \cap B(x_3, r_3) = \emptyset. \tag{6.3}$$

To obtain a triple intersection, one must increase the radii. Thus, we seek the smallest $\lambda \geq 1$ such that:

$$B(x_1, \lambda r_1) \cap B(x_2, \lambda r_2) \cap B(x_3, \lambda r_3) \neq \emptyset. \tag{6.4}$$

[7] For a more detailed mathematical analysis, see Barth et al. (2025), which addresses some shortcomings and introduces an improved method, IsUMap (Barth et al. 2025).

A characteristic property of Riemannian sectional curvature is that the smaller the curvature, the smaller the required λ. This property can be used, analogously to the Alexandrov procedure described earlier, to define curvature bounds for general metric spaces.

If the spaces are discrete, a small correction term must be allowed for. Instead of a radius λr, one considers a corrected radius $\lambda r + \delta$, where $\delta > 0$ reflects the scale at which the discreteness of the space becomes significant. For large radii r, a small δ is, of course, negligible.

Such constructions are therefore useful for capturing the global structure of metric spaces.

References

Aleksandrov, Aleksander D., Valerii N. Berestovskii, and Igor G. Nikolaev. 1986. Generalized Riemannian spaces. *Russian Mathematical Surveys* 41(3): 1.

Alexandrov, Aleksander Danilovich. 1957. Über eine Verallgemeinerung der Riemannschen Geometrie. *Schr. Forschungsinst. Math. Berlin* 1: 33–84.

Barth, Lukas Silvester, Fatemeh Fahimi, Parvaneh Joharinad, Jürgen Jost, and Janis Keck. 2025. *Data Visualization with Category Theory and Geometry*. Math of Data. Belin: Springer.

Barth, Lukas Silvester, Fatemeh Fahimi, Parvaneh Joharinad, Jürgen Jost, and Janis Keck. 2025. *Isumap: Manifold Learning and Data Visualization Leveraging Vietoris-Rips Filtrations*. In *Proceedings of the AAAI Conference on Artificial Intelligence* (Vol. 39, No. 17, pp. 17699–17706).

Belkin, Mikhail, and Partha Niyogi. 2003. Laplacian eigenmaps for dimensionality reduction and data representation. *Neural Computation* 15 (6): 1373–1396.

Busemann, Herbert. 1955. *The Geometry of Geodesics*. New York: Academic Press.

Burago, Dmitri, Yu Burago, and Sergei Ivanov. *A course in metric geometry*. AMS, 2001.

Freudenthal, Hans, ed. 1978. *Raumtheorie*. Darmstadt: Wiss. Buchges.

Gromov, Mikhail. 1980. *Structures métriques pour les variétés riemanniennes*. ed. J. Lafontaine, and P. Pansu. Paris: Cedic-Nathan.

Gromov, Mikhail. 1999. *Metric Structures for Riemannian and Non-Riemannian Spaces*. Birkhäuser.

Hinton, Geoffrey E., and Sam Roweis. 2002. Stochastic neighbor embedding. *Advances in Neural Information Processing Systems* 15.

Joharinad, Parvaneh, and Jürgen Jost. 2019. Topology and curvature of metric spaces. *Advances in Mathematics* 356: 106813.

Joharinad, Parvaneh, and Jürgen. Jost. 2023. *Mathematical Principles of Topological and Geometric Data Analysis*. Mathematics of Data. Berlin: Springer.

Jost, Jürgen. 1997. *Nonpositive Curvature: Geometric and Analytic Aspects*. Birkhäuser.

Jost, Jürgen. 2017. *Riemannian geometry and geometric analysis*, 7th edn. Berlin: Springer.

Jost, Jürgen. 2026. *Riemannian geometry and geometric analysis*, 8th edn. Berlin: Springer.

McInnes, Leland, John Healy, and James Melville. 2018. Umap: Uniform manifold approximation and projection for dimension reduction. https://arxiv.org/abs/1802.03426.

Menger, Karl. 1929. Untersuchungen über allgemeine Metriken. Vierte Untersuchung. Zur Metrik der Kurven. *Mathematische Annalen* 103: 466–501.

Wald, Abraham. 1936. Begründung einer koordinatenlosen Differentialgeometrie der Flächen. *Ergebnisse eines Math.Kolloquiums, 1.Reihe* 7: 24–46.

Selected Bibliography with Commentaries 7

This bibliography is not intended to be complete. Literature concerning special aspects is listed in the footnotes and at the end of a chapter. A superscript in front of a year denotes the number of the edition; for instance, ^{2}1990 means "2nd ed., 1990".

7.1 Different Editions of the Text

The original source is Riemann' s habilitation lecture on June 10, 1854. It was published only after Riemann's death by Richard Dedekind in 1868:

Bernhard Riemann, *Über die Hypothesen, welche der Geometrie zu Grunde liegen.* (Aus dem Nachlaß des Verfassers mitgetheilt durch R. Dedekind). Abh. Ges. Gött., Math. Kl. 13 (1868), 133–152.

Reprinted in

Bernhard Riemann's gesammelte mathematische Werke und wissenschaftlicher Nachlass. Herausgegeben unter Mitwirkung von Richard Dedekind von Heinrich Weber, 1. Aufl., Leipzig, Teubner-Verlag, 1876, 254–269; 2. Aufl. bearbeitet von Heinrich Weber, Leipzig, Teubner-Verlag, 1892, 272–287.

On the basis of the edition of collected works from 1892 and the supplements from 1902 (Bernhard Riemann, *Gesammelte mathematische Werke. Nachträge.* Herausgegeben von M. Noether und W. Wirtinger. Leipzig, Teubner-Verlag, 1902), there are the more recent editions

Bernhard Riemann, *Collected works*, with a new introduction by Hans Lewy, New York, Dover, 1953.

Bernhard Riemann, *Gesammelte mathematische Werke und wissenschaftlicher Nachlass und Nachträge. Collected Papers.* Nach der Ausgabe von Heinrich Weber und Richard Dedekind neu herausgegeben von Raghavan Narasimhan, Berlin etc., Springer-Verlag, and

© The Author(s), under exclusive license to Springer Nature Switzerland AG 2025
J. Jost, *Bernhard Riemann — On the Hypotheses Which Lie at the Bases of Geometry*,
Classic Texts in the Sciences, https://doi.org/10.1007/978-3-031-91489-8_7

Leipzig, Teubner-Verlag, 1990, 304–319 (this edition has a double pagination, in addition to the sequential one also a reproduction of the Weber-Dedekind edition from 1892).

Riemann's habilitation lecture is also reproduced as Bernhard Riemann. *Über die Hypothesen, welche der Geometrie zu Grunde liegen.* Neu herausgegeben und erläutert von H. Weyl, Berlin, Springer-Verlag, 1919, ^{3}1923.

This edition with Weyl's commentary is reproduced in turn in: *Das Kontinuum und andere Monographien*, New York, Chelsea Publ. Comp., 1960, ^{2}1973. The preface and the commentary by Hermann Weyl are also reprinted in the Narasimhan edition, 740–768.

C. F. Gauß/B. Riemann/H. Minkowski, *Gaußsche Flächentheorie, Riemannsche Räume und Minkowskiwelt.* Herausgegeben und mit einem Anhang versehen von J. Böhm und H. Reichardt, Leipzig, Teubner-Verlag, 1984, 68–83.

Translations can be found in

Bernhard Riemann, *Œuvres mathématiques*, traduites par L. Langel, avec une préface du M. Hermite et un discours de M. Félix Klein, Gauthier-Villard, Paris, 1898, reprinted by Ed. Jacques Gabay, Paris, 1990, 2003, also available from Univ. Michigan Press, 2006.

William Kingdon Clifford (1845-1879) in *Nature*, Vol. VIII, Nos. 183,184, 1873, pp. 14–17,36,37; reproduced in W. Clifford, *Mathematical papers*, edited by Robert Tucker, with an introduction by H. J. Stephen Smith, London, MacMillan and Co., 1882, pp. 55–71 (this translation is reproduced here).

David E. Smith, *A source book in mathematics,* McGraw-Hill, 1929, and Mineola, N. Y., Dover, 1959, 411–425.

Michael Spivak, *A comprehensive introduction to differential geometry*, Vol. 2, Berkeley, Publish or Perish, 1970 (with commentary).

Riemann's article on heat diffusion, Commentatio mathematica, qua respondere tentatur quaestioni ab Illma Academia Parisiensi propositae: "Trouver quel doit être l'état calorifique d'un corps solide homogène indéfini pour qu' un système de courbes isothermes, à un instant donné, restent isothermes après un temps quelconque, de telle sorte que la température d'un point puisse s'exprimer en fonction du temps et de deux autres variables indépendantes", in which Riemann translates his geometric concepts into a mathematical formalism, can be found in *Gesammelte Werke*, 2. Aufl., 423–436, with extensive commentaries by the editors, ibd. 437–455 (according to Narasimhan's pagination). There is a German translation of the Latin text by O. Neumann in the volume edited by Böhm and Reichardt, p. 115–128. A partial translation with a detailed commentary is in Spivak, Vol. 2.[1]

Detailed commentaries are in

Alberto Cogliati, *Riemann's commentatio mathematica, a reassessment.* Revue d'histoire des mathématiques 20.1, 2014, 73–94.

Olivier Darrigol, *The mystery of Riemann's curvature*. Historia mathematica 42, 2015, 47–83.

The texts of Helmholtz appeared originally as

[1] As Spivak writes in his preface "the fact that I don't know Latin didn't hinder me much", you should not expect high philological accuracy.

Hermann Helmholtz, *Ueber die thatsächlichen Grundlagen der Geometrie,* Verhandlungen des naturhistorisch-medicinischenVereins zu Heidelberg, Bd. IV, 197–202, 1868; Zusatz ebd. Bd. V, 31–32, 1869.

Hermann Helmholtz, *Ueber die Thatsachen, die der Geometrie zu Grunde liegen,* Nachrichten der Königl. Gesellschaft der Wissenschaften zu Göttingen 9, 193–221, 1868. quoted here after

Hermann Helmholtz, *Wissenschaftliche Abhandlungen*, Bd. 2, Leipzig, Johann Ambrosius Barth, 1883.[2]

Moreover

Hermann Helmholtz, *Ueber den Ursprung und Sinn der geometrischen Axiome,* in: *Populäre wissenschaftliche Vorträge,* Heft III, 21–54, and in ders., *Vorträge und Reden,* Bd. II, Braunschweig, 1–31, 1884, which I quote after

Hermann von Helmholtz, *Schriften zur Erkenntnistheorie.* Kommentiert von Moritz Schlick und Paul Hertz. Herausgegeben von Ecke Bonk, Wien/New York, Springer, 1998, which is the reprint of the edition on the occasion of the 100th anniversary of his birth, Berlin, Springer, 1921. Another new edition is

Hermann von Helmholtz, *Schriften zur Erkenntnistheorie.* Herausgegeben von Moritz Schlick und Paul Hertz, Saarbrücken, Dr. Müller, 2006.

Furthermore

Hermann Helmholtz, *Ueber den Ursprung und Sinn der geometrischen Sätze; Antwort gegen Herrn Professor Land,* in his *Wiss. Abh.*, Vol. II (An English translation appeared in Mind 3, 212–225, 1878).
also reproduced together with his other writings on the subject in

Hermann von Helmholtz, *Ueber Geometrie*, Darmstadt, Wiss. Buchges., 1968.

A slightly shortened version can also be found in the appendix of

Hermann von Helmholtz, *Die Thatsachen in der Wahrnehmung*, Berlin, A. Hirschwald, 1879, which in turn is reproduced in his *Schriften zur Erkenntnistheorie*.

Hermann Helmholtz, *Gesammelte Schriften,* mit einer Einleitung herausgegeben von Jochen Brüning, 7 Vols. in 19 Subvols., Hildesheim, Olms, 2001ff

is not yet complete.

7.2 Bibliographies

There is an extensive bibliography on Riemann, compiled by W. Purkert and E. Neuenschwander, in Narasimhan's edition of the collected works. The mathematical research

[2] On p. 610, however, the year of publication of the *thatsächlichen Grundlagen* seems incorrect, 1866 instead of 1868. In particular, Helmholtz mentions there, on p. 611, already the publication of Riemann's text, which took place only in 1868.

articles on Riemannian geometry are far too numerous to list them in a bibliography. Newer contributions are collected in the preprint server http://arXiv.org in the category *Differential Geometry*.

7.3 Introductions

The essential source for Riemann's life remains the biography written by Dedekind for the collected works. Some further biographical details can be found in

Erwin Neuenschwander, *Lettres de Bernhard Riemann à sa famille,* Cahiers du Séminaire d'Histoire des Mathématiques 2, 85–131, 1981.

Scientific biographies of Riemann are

Felix Klein, *Riemann and his significance for the development of modern mathematics,* Bull. Amer. Math. Soc. 1, no. 7, 165–180, 1895 (translated from the German *Riemann und seine Bedeutung für die Entwicklung der modernen Mathematik*, Jahresber. Deutsche Mathematiker-Vereinigung 4, 71–87, 1894/95, reprinted in the same, *Gesammelte mathematische Abhandlungen*, Bd. 3, 482–497, Berlin, Springer, 1923).

Hans Freudenthal, *Riemann, Georg Friedrich Bernhard*, Dictionary of Scientific Biography, Vol. 11, New York, 447–456.

Lizhen Ji, Shing Tung Yau, *What one should know about Riemann but may not know?*, in: L.Z.Ji, S.T.Yau (eds.), *The Legacy of Bernhard Riemann After One Hundred and Fifty Years*, Volume I, pp. 1–56, International Press, 2016.

Detlef Laugwitz, *Bernhard Riemann 1826–1866. Turning points in the conception of mathematics* (translated from the German), Boston, Birkhäuser, 2008.

Michael Monastyrsky, *Riemann, topology, and physics,* Boston etc., Birkhäuser, ^{3}2008.

The influence of Riemann's ideas is also discussed in

Krysztof Maurin, *The Riemann legacy. Riemannian ideas in mathematics and physics of the 20*th *century*, Dordrecht, Kluwer, 1997.

Lizhen Ji, Shing Tung Yau (eds.), *The Legacy of Bernhard Riemann After One Hundred and Fifty Years*, 2 Volumes, International Press, 2016.

Among treatises on the history of mathematics, the still most important one for our topic remains

Felix Klein, *Development of mathematics in the 19th century,* with a preface and appendices by Robert Hermann. (Translated by M. Ackerman from the German *Vorlesungen über die Entwicklung der Mathematik im 19. Jahrhundert,* 2 Vols., Berlin, Springer, 1926/7, reprinted as a single vol., Berlin etc., Springer, 1979.) Lie Groups: History, Frontiers and Applications, IX. Math. Sci. Press, Brookline, Mass., 1979. About Riemann, see in particular pp. 175–180. Klein was not only an important mathematician himself, but he was also in the position to present the development of mathematics from his personal acquaintance with most of the key protagonists.

For a general and comprehensive history of mathematical thinking, we refer to

Morris Kline, *Mathematical thought. From ancient to modern times*, 3 vols., Oxford, Oxford Univ. Press, ^{2}1990.

From a group of authors:

Jean Dieudonné, *Abrégé d'histoire des mathématiques: 1700–1900*, revised ed., Editons Hermann, Paris, 1996 (German translation of the original edition, Paris, Hermann, 1978: *Geschichte der Mathematik 1700–1900. Ein Abriß*, Braunschweig, Wiesbaden, Vieweg, 1985); in particular Paulette Libermann, Chap. 9: Géometrie differentielle.

On non-Euclidean geometry, with translations of the original articles on non-Euclidean geometry by Bolyai and Lobachevski

Roberto Bonola, *Non-Euclidean geometry. A critical and historical study of its development,* translated from the Italian and with additional appendices by H. S. Carslaw. With an introduction by Federico Enriques. With a supplement containing the Dr.George Bruce Halsted translations of *The science of absolute space* by John Bolyai and *The theory of parallels* by Nicholas Lobachevski, New York, Dover, 1955.

Several aspects of non-Euclidean geometry were anticipated by earlier mathematicians, in particular Gerolamo Saccheri and Johann Heinrich Lambert, see

Gerolamo Saccheri, *Euclid vindicated from every blemish*, edited and annotated by Vincenzo De Risi. Translated by G. B. Halsted and L. Allegri. In: Classic Texts in the Sciences (O. Breidbach, J. Jost, eds.), Basel, Birkhäuser, 2014.

and

Johann Heinrich Lambert, *Theorie der Parallellinien*, edited and annotated by Vincenzo De Risi. In: Classic Texts in the Sciences (O. Breidbach, J. Jost, eds.), Basel, Birkhäuser, to appear.

Further detailed bibliographical references are for example given in

Felix Klein, *Vorlesungen über nicht-euklidische Geometrie*, Berlin, Springer, 1928, in particular p. 275f., and concerning the role of Riemannian geometry, pp. 288–293.

Some newer presentations of this topic:

Jeremy J. Gray, *Ideas of Space. Euclidean, Non-Euclidean, and Relativistic.* Oxford Univ. Press, ^{2}1989.

Jeremy J. Gray, *Worlds Out of Nothing.* A Course in the History of Geometry in the nineteenth Century. Berlin etc., Springer, 2007.

An introduction to the theory of relativity by its creator:

Albert Einstein, *Relativity. The special and general theory,* translated from the German by R. Lawson, New York, Henry Holt, 1920; various subsequent editions and reprints.

The history of ideas aspects are treated in Oskar Becker, *Grundlagen der Mathematik,* loc.cit.

This leads us to some works that analyze the problem of space from the perspective of the history of ideas. A fundamental treatise is

Max Jammer, *Concepts of space: History of theories of space in physics,* Cambridge MA., Harvard Univ. Press, ^{2}1980.

A wealth of material is contained in

Alexander Gosztonyi, *Der Raum. Geschichte seiner Probleme in Philosophie und Wissenschaften,* 2 Vols., Freiburg, München, Karl Alber, 1976.

A more recent work that develops the relevant physical ideas and the different natural philosophical positions is

Martin Carrier, *Raum-Zeit*, Berlin, de Gruyter, 2009.

On the philosophy of mathematics, we quote the fundamental monograph

Hermann Weyl, *Philosophy of mathematics and natural science*, translated from the German by O. Helmer, Princeton, Princeton Univ. Press, ^{2}2009.

as well as

Léon Brunschvicg, *Les étapes de la philosophie mathématique*, Paris, Presses Univ. France, ^{3}1947

Roberto Torretti, *The philosophy of physics*, Cambridge, Cambridge Univ. Press, 1999, pp. 157–168, contains a thorough discussion of Riemann's *Hypothesen*. Riemann is also treated in detail in

Helmut Pulte, *Axiomatik und Empirie.* Eine wissenschaftstheoriegeschichtliche Untersuchung zur Mathematischen Naturphilosophie von Newton bis Neumann. Darmstadt, Wiss. Buchges., 2005, pp. 359–401.

We also mention

Peter Mittelstaedt, *Philosophische Probleme der modernen Physik*, Mannheim, Bibliograph. Inst, ^{2}1966.

7.4 Important Monographs and Articles

On issues concerning the history of mathematics and philosophy

Luciano Boi, *Le problème mathématique de l'espace*, Berlin, Heidelberg, Springer, 1995.

Joël Merker, *Sophus Lie, Friedrich Engel, et le problème de Riemann-Helmholtz*, arXiv:0910.0801v1, 2009, a French translation with commentary of the *Theorie der Transformationsgruppen* (Dritter und letzter Abschnitt, Abtheilung V) of Lie and Engel with a detailed treatment of the considerations of Riemann and Helmholtz.

Karin Reich, *Die Geschichte der Differentialgeometrie von Gauß bis Riemann (1828–1868),* Archive for History of Exact Sciences 11, 273–382, 1973.

Erhard Scholz, *Geschichte des Mannigfaltigkeitsbegriffs von Riemann bis Poincaré,* Boston etc., Birkhäuser, 1980.

Erhard Scholz, *Herbart's influence on Bernhard Riemann*, Historia Mathematica 9, 413–440, 1982.

Erhard Scholz, *Riemanns frühe Notizen zum Mannigfaltigkeitsbegriff und zu den Grundlagen der Geometrie*, Archive for History of Exact Sciences 27, 213–232, 1982.

Andreas Speiser, *Naturphilosophische Untersuchungen von Euler und Riemann*, Journal für die reine und angewandte Mathematik 157, 105–114, 1927.

Roberto Torretti, *Philosophy of geometry from Riemann to Poincaré,* Dordrecht etc., Reidel, 1978.

André Weil, *Riemann, Betti and the birth of topology*, Archive for History of Exact Sciences 20, 91–96, 1979; Postscript in Archive for History of Exact Sciences 21, 387, 1980.

On the theory of general relativity and its mathematical penetration and its influence on the development of geometry

Albert Einstein, *Die Feldgleichungen der Gravitation*, Sitzungsber. Preußische Akademie der Wissenschaften 1915, 844–847.

David Hilbert, *Die Grundlagen der Physik*, Königl. Gesellschaft der Wissenschaften Göttingen, Mathematisch-Physikalische Klasse, 395–407, 1915; 53–76, 1917; a revised version is reprinted in

David Hilbert, *Die Grundlagen der Physik*, Math. Annalen 92, 1–32, 1924, and in

David Hilbert, *Gesammelte Abhandlungen,* Bd. III, Berlin etc., Springer, ^{2}1970, S. 258–289.

Albert Einstein, *Die Grundlage der allgemeinen Relativitätstheorie*, Annalen der Physik 49, 769–822, 1916.

The articles of Einstein on the theory of relativity are reprinted in

Albert Einsteins Relativitätstheorie. Die grundlegenden Arbeiten. Herausgegeben und erläutert von Karl von Meyenn, Braunschweig, Vieweg, 1990.

and with a detailed analysis of the development of Einstein's theory in

Michel Janssen and Jürgen Renn, *How Einstein Found His Field Equations: Sources and Interpretation*, Classic Texts in the Sciences, Springer Nature, 2022.

Hermann Weyl, *Space, time, matter,* translated from the German, revised ed., Mineola NY, Dover, 1952 (the translation is not very good; a more recent version of the German original is *Raum, Zeit, Materie*, ed. by Jürgen Ehlers, Berlin, Springer, ^{7}1988).

and with a detailed commentary

Raum, Zeit, Materie, ed. by Domenico Giulini and Erhard Scholz, based on the edition of 1923, Klassische Texte der Wissenschaft, Springer, 2025.

Hermann Weyl, *Mathematische Analyse des Raumproblems*, Berlin, Springer, 1923.

Charles Misner, Kip Thorne and John Archibald Wheeler, *Gravitation*, New York, Freeman, 1973.

There exists an extensive literature on the history and the impact of the general theory of relativity. Here, we only cite the collection

Jürgen Renn (ed.), *The Genesis of General Relativity. Sources and Interpretations.* 4 Bde. Berlin etc., Springer, 2007.

with its detailed commentaries.

On the current state of research in geometry and theoretical physics

Marcel Berger, *A panoramic view of Riemannian geometry*, Berlin etc., Springer, 2003

Pierre Deligne et al. (eds.), *Quantum fields and strings: A course for mathematicians,* 2 Vols., Princeton, Amer. Math. Soc., 1999.

M. B. Green, J. H. Schwarz und E. Witten, *Superstring theory*, 2 Vols., Cambridge etc., Cambridge Univ. Press, 1987.

S. W. Hawking und G. F. R. Ellis, *The large scale structure of space-time*, Cambridge etc., Cambridge Univ. Press, 1973.

Sigurdur Helgason, *Differential geometry, Lie groups, and symmetric spaces*, New York etc., Academic Press, 1978.

Jürgen Jost, *Riemannian geometry and geometric analysis*, Berlin etc., Springer, ^{7}2017, ^{8}2026.

Jürgen Jost, *Geometry and physics*, Berlin etc., Springer, 2009.

Jürgen Jost, *Mathematical concepts*, Berlin etc., Springer, 2015.

Wilhelm Klingenberg, *Riemannian geometry*, Berlin, New York, de Gruyter, 1982.

Roger Penrose, *The road to reality. A complete guide to the laws of the universe*, London, Jonathan Cape, 2004.

Steven Weinberg, *The quantum theory of fields,* 3 vols., Cambridge etc., Cambridge Univ. Press, 1995, 1996, 2000.

Eberhard Zeidler, *Quantum field theory*, 3 vols., Berlin etc., Springer, 2006 ff.

Biographical Outline and Chronological Table

The historical events of the Napoleonic wars and the establishment of the German Reich frame the life span of Riemann and the aftermath of the first and the preparations for the second shaped the political and economic situation of the time, in which Riemann lived.[1] Of obvious importance for the understanding of the scientific development and the life of Riemann was the situation in the German universities, especially Göttingen and Berlin, and of course the general development of mathematics. This will be reflected in the following chronological table.

1737: Opening of the University of Göttingen, where the role of science is highlighted.
1801: Carl Friedrich Gauss' "Disquisitiones arithmeticae" appear.
1806: Formal end of the Holy Roman Empire of the German Nation, as Emperor Francis II. under the pressure of Napoleon lays down the German imperial crown. Collapse of Prussia after the battle of Jena and Auerstädt. Napoleon enters Berlin.
1807: In response to the inferiority of Prussia against Napoleon's aggression introduction of far-reaching reforms in Prussia by the Baron vom Stein.
1810: In the wake of these reforms Wilhelm von Humboldt initiates the founding of the University of Berlin. His university constitution will become decisive for the academic life in the nineteenth century in Germany.
1813: Beginning of the wars of liberation against France, in which also Riemann's father participates. Napoleon's defeat in the Battle of the Nations near Leipzig.
1815: Final defeat of Napoleon and reorganization of Europe in the Congress of Vienna. Founding of the German Confederation. The beginning of the Restoration period shaped by the Austrian Chancellor Metternich.

[1] The following facts about Riemann's life are mostly taken from his biography written by Dedekind in the Collected Works of Riemann. I have also used Laugwitz, *Riemann*. However, I did not check the original sources systematically.

© The Editor(s) (if applicable) and The Author(s), under exclusive license to Springer Nature Switzerland AG 2025

J. Jost, *Bernhard Riemann — On the Hypotheses Which Lie at the Bases of Geometry*, Classic Texts in the Sciences, https://doi.org/10.1007/978-3-031-91489-8

1817: Establishment of the Ministry of Culture in Prussia, with Altenstein as its the first and longtime head.

1818: Prussian Customs Act creates the conditions for the advancement of Prussia as a leading economic power.

1819: Kingdom of Hanover receives Constitution.

1820: Vienna Final Act completed constitution of the German Confederation.

1826: Birth of Georg Friedrich Bernhard Riemann on September 17 as the eldest son of the local Protestant minister in Breselenz near Dannenberg on the river Elbe in the Kingdom of Hanover. Childhood in nearby Quickborn in the Elbe lowland, where the father becomes head of the Parish and teaches his children.

1827: Carl Friedrich Gauss' "Disquisitiones generales circa superficies curvas" create modern differential geometry.

1831: Student riots in Göttingen. Death of Hegel, which marks the end of the height of German idealism. Faraday discovers the electromagnetic induction.

1832: Goethe dies, bringing the Weimar Classicism to its end.

1834: German Zollverein under Prussian leadership. Death of Schleiermacher, the founder of modern Protestant theology. Jacobi founds in Königsberg the first Mathematical and Physical Seminar in Germany.

1837: End of the personal union between Hanover and Great Britain, since Hanover does not allow the female succession of the British Queen Victoria. The new Hanoverian King Ernest Augustus drives a reactionary turn. Dismissal of the "Göttingen Seven" which include the physicist Wilhelm Weber who had collaborated with Gauss, because of their protest against the violation of the Constitution.

1840: Frederick William IV King of Prussia. He disappoints the expectations placed on him for a liberal policy and pursues instead a conservative-reactionary course. Riemann visits Hanover high school (until 1842) and lives there with his Grandmother.

1841: Death of the architect Schinkel, who had built Berlin on behalf of the Prussian Royal Family as a modern European capital with recourse on many architectural styles.

1842: After the death of his grandmother Riemann attends Gymnasium in Lüneburg (until 1846), whose director Schmalfuss recognizes and promotes Riemann's great mathematical talent.

1846: Death of Riemann's mother. Riemann begins his studies at the University of Göttingen, on his father's wish first of theology, but soon switches to mathematics.

1847: Riemann moves to the University of Berlin and attends the Lectures of Dirichlet and Jacobi and gets in contact with Eisenstein, which, however, for personal reasons is not very fertile.

1848: The "Communist Manifesto" by Marx and Engels appears. Beginning of revolutions in various countries, notably France, Austria (fall of Metternich) and Prussia. Frankfurt National Assembly in St. Paul's Church. The annexation of Schleswig by Denmark leads to the 1st German-Danish War. The Prussian National Assembly is dissolved. Election of Louis Napoleon President of France.

1849: The Prussian King Friedrich Wilhelm IV rejects the imperial crown of Lesser Germany offered by the Frankfurt National Assembly. Suppression of uprisings in support of the constitution in various German states. Riemann is an eyewitness to the March Revolution in Prussia and accepts a short guard duty as a member of the student corps. Dissolution of the National Assembly. Riemann returns to the University Göttingen, where Weber regains his physics professorship and also promotes Riemann personally.

1850: Under pressure from Austria, Friedrich Wilhelm IV abandons his efforts concerning a new constitution for Germany. A Prussian Constitution comes into effect. Clausius formulates the Second Law of Thermodynamics. Riemann enters into the recently founded mathematical-physical seminar in Göttingen. Dedekind begins his studies in Göttingen and becomes Riemann's lifelong friend.

1851: PhD of Riemann with Gauss.

1852: Louis Napoléon becomes the French Emperor Napoleon III. Dirichlet visits Göttingen in the fall; many scientific discussions with Riemann.

1854: Habilitation of Riemann at the Faculty of Philosophy of the University of Göttingen; Habilitation colloquium on June 10 on the subject "Ueber die Hypothesen, welche der Geometrie zu Grunde liegen".

1855: Riemann's father and one of his four sisters die. Death of Gauss and appointment of Dirichlet as his successor in Göttingen.

1856: Heine dies in Paris.

1857: Riemann becomes an associate professor in Göttingen. Death of Riemann's younger brother Wilhelm. Riemann takes over the financial care of his three surviving sisters. The work "Theorie der Abelschen Funktionen" gives Riemann high scientific recognition.

1858: Prince William takes over the government in Prussia for his incapacitated brother Frederic William IV who is declared as unfit for government. Riemann meets the Italian mathematicians Brioschi, Betti and Casaroti, who visit Göttingen. Dedekind accepts a chair at the Polytechnic in Zürich and leaves Göttingen.

1859: "On the Origin of Species" by Darwin founds modern evolutionary biology. Death of Alexander von Humboldt, the most prominent German scientist of his time. Death of Dirichlet and appointment of Riemann as his successor as chaired professor. Riemann corresponding member of the Bavarian Academy and the Berlin Academy; travel to Berlin, accompanied by Dedekind. Riemann full member of the Society of Sciences in Göttingen.

1860: Risorgimento and the Unification of Italy under the leadership of Piedmont. Riemann visits Paris for a month and comes into contact with Parisian mathematicians.

1861: Proclamation of the Kingdom of Italy. Due to an amnesty Wagner can return to Germany, which he had had to leave for his participation in the failed revolution of 1849 in Saxony.

1862: Bismarck appointed Prime Minister of Prussia. Marriage of Riemann with Elise Koch. A pleurisy causes permanent damage to his lungs. First trip to Italy, in the hope that the local mild climate is beneficial for his health.

1863: On the way back to Göttingen Riemann gets in close contact with the mathematician Enrico Betti in Pisa. After two months in Göttingen new trip to Italy, where his daughter Ida is born. Declines a position at the University of Pisa. Riemann becomes a full member of the Bavarian Academy.

1864: 2nd German-Danish War. Maxwell formulates his theory of electromagnetism.

1866: Prussia attains German supremacy after victory in the war against Austria. Dissolution of the German Confederation. Prussia takes over the Kingdom of Hanover. Riemann becomes a foreign member of the Paris Academy and the Royal Society in London. During the first days of the war, he goes on a new journey to Italy. Riemann dies on July 20 in Selasca at Lake Maggiore.

1867: Creation of the North German Confederation with Prussia as a hegemonic power.

1868: Posthumous publication of Riemann's habilitation lecture on the initiative of Dedekind. Helmholtz's "Ueber die Thatsachen, die der Geometrie zu Grunde liegen" appears.

1871: After Prussia's victory against France establishment of the German Empire under Prussian leadership.

1876: The collected works of Riemann appear.

1884: Riemann's daughter Ida (1863-1929) marries Carl David Schilling (1857-1932) who had obtained his Ph.D. with Hermann Amandus Schwarz (1843-1921) in Göttingen and will later become Director of the Seefahrtsschule in Bremen whence in 1890 also Riemann's wife and his only surviving sister Ida will move. The couple will have 5 children.

1892: 2nd edition of the collected works of Riemann.

1907–1916: Einstein is working on the general theory of relativity.

1990: New edition of the collected works of Riemann.

Glossary

Manifold Term for the continuous juxtaposition of points or elements, provided that sufficiently small parts can be mapped bijectively, i.e. in an invertible manner, to a portion of the Cartesian space by a tuple of numbers, the **coordinates**. The concept of a **manifold** is purely topological, in the sense that it does not presuppose a **metric structure**, and involves therefore only qualitative relations situation. Although a spatial concept, the space imagined does not need to be the physical space. For instance, the different color values constitute the elements of a manifold, the color space.

Coordinates Representation of a portion of a **manifold** by a domain in a Cartesian space. The position of a point in an n-dimensional Cartesian space is specified by n real numbers. These n numbers are then called the **coordinates** of the point on the **manifold** corresponding to this point in the Cartesian space. **Coordinates** thus provide the possibility to describe the position of a point in a manifold by real numbers. This description or specification of a point in a **manifold** is, however, not inherent in the point, but only a convention. In different **coordinates**, one and the same point is described by different numbers.

Dimension How many real numbers are required to represent each point in a **manifold** uniquely by **coordinates**.

Metric Determination of the **distances** between the points of a **manifold** (or of a more general metric space); axiomatically given mathematical structure formulating the conditions for the notion of a **distance** (any two distinct points must always have a positive **distance** from each other, which does not depend on the order of the two points. and the **triangle inequality** holds, i.e. that the **distance** of two points from each other cannot be larger than the sum of their distances from a third point).

Riemannian metric Quadratic form on a **manifold**, which allows for the computation of the length of curves by integration along them. Moreover, when lengths can be assigned to curves, the distance between two points is the smallest length among all the curves connecting them. More precisely, we should speak of the quadratic form defining the

© The Editor(s) (if applicable) and The Author(s), under exclusive license to Springer Nature Switzerland AG 2025

J. Jost, *Bernhard Riemann — On the Hypotheses Which Lie at the Bases of Geometry*, Classic Texts in the Sciences, https://doi.org/10.1007/978-3-031-91489-8

metric , since the latter is an infinitesimal notion in distinction to the **distance** notion that yields the **metric**.

Riemannian manifold A **manifold** equipped with a Riemannian **metric**.

Curvature Measure for the deviation of a surface, or more generally, a **manifold** from a flat, Euclidean shape.

Invariant Quantity that does not change under a class of **transformations** or that remains the same in different descriptions. For instance, the **dimension** or the **curvature** of a **manifold** do not depend on the choice of **coordinates** and are hence coordinate invariant.

Surface theory Theory for the description of two-dimensional objects.

Non-Euclidean geometry Space structure in which the parallel postulate does not hold, but all the other Euclidean **axioms** hold.

Parallel transport Transport of direction elements (tangent vectors) from one point of a V Riemannian **manifold** to another one along some curve, such that their lengths and the angles between them remain invariant.

Topology Theory of the qualitative relations between the points of a mathematical space. Metric relations, however, are of a quantitative nature and therefore do not fall into the realm of topology.

Index

© The Editor(s) (if applicable) and The Author(s), under exclusive license to Springer Nature Switzerland AG 2025

J. Jost, *Bernhard Riemann — On the Hypotheses Which Lie at the Bases of Geometry*, Classic Texts in the Sciences, https://doi.org/10.1007/978-3-031-91489-8

The manufacturer's authorised representative in the EU is Springer
Nature Customer Service Centre GmbH, Europaplatz 3, 69115 Heidelberg,
Germany. If you have any concerns regarding our products, please
contact ProductSafety@springernature.com

Printed and bound by CPI Group (UK) Ltd, Croydon, CR0 4YY
30/09/2026
02225507-0020